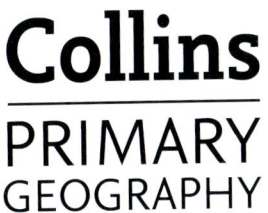

Change
Pupil Book 5

Planet Earth
Seas and oceans
Beneath the surface — 2
The ocean environment — 4
Learning about seas — 6

Water
Wearing away the land
Rivers in action — 8
Preventing flood damage — 10
Finding out about rivers — 12

Weather
The seasons
Changing seasons — 14
Seasons worldwide — 16
Seasonal influences — 18

Settlements
Cities
Describing cities — 20
World cities — 22
The story of London — 24

Work and Travel
Jobs
Making things — 26
Different jobs — 28
Types of work — 30

Environment
Pollution
Damaging the environment — 32
'Green living' — 34
Exploring clean energy — 36

Places
Wales — 38
Greece — 44
North America — 50
Africa — 56

Glossary — 62

Index — 63

Stephen Scoffham | Colin Bridge

Unit 1 — Seas and oceans

Lesson 1: Beneath the surface

What is it like under the oceans?

We know less about the oceans than any other part of the world. People want to find out more about the animals that live in the water, how the oceans affect the weather and what happens on the ocean floor.

Exploring the oceans is difficult. There is plenty of light at the surface but below 200 metres, it is almost completely dark. The weight of water is so heavy that people can only survive if they are in a submarine. It is also very cold.

Scientists have now discovered underwater vents. These pump fountains of boiling water and minerals into the ocean. Large numbers of animals live around the vents. Some of the animals have shells and look like crabs and shrimps. There are also huge worms that have no mouths or stomachs.

In many places the ocean floor is several kilometres below the surface.

Discussion
- Why do people want to explore the oceans?
- What makes exploring the oceans difficult?
- What is the difference between deep-sea creatures and those that live near the surface?

Key words
coral reef
minerals
ocean floor
trench
vent
volcano

Unit 1   Seas and oceans

Data bank
- There are more volcanoes under the ocean than on dry land.
- The Marianas Trench in the Pacific Ocean is so deep (nearly 11 000 metres) that Mount Everest would fit into it.

▲ These beautiful but poisonous sea anemones live on the Great Barrier Reef, Australia.

▲ This ancient shell creature (a nautilus) normally lives 150 to 300 metres below the surface.

▲ Below 500 metres there are unusual fish, like the Northern wolffish.

Mapwork
Using an atlas and blank map of the world, show some of the different places where coral reefs are found.

Investigation
Make a class scrapbook about the oceans using online research, newspapers and magazines.

Unit 1 Seas and oceans

Lesson 2: The ocean environment

What are the threats to the ocean environment?

Key words
climate
equator
global warming
ocean currents
Tropics

Climate change

Global warming is causing seas and oceans to get warmer, so more water evaporates. This makes storms more powerful, especially in the Tropics. The warmer water is also bad for creatures and coral reefs. These are home to a quarter of all sea life.

▼ Storm clouds gather over the ocean.

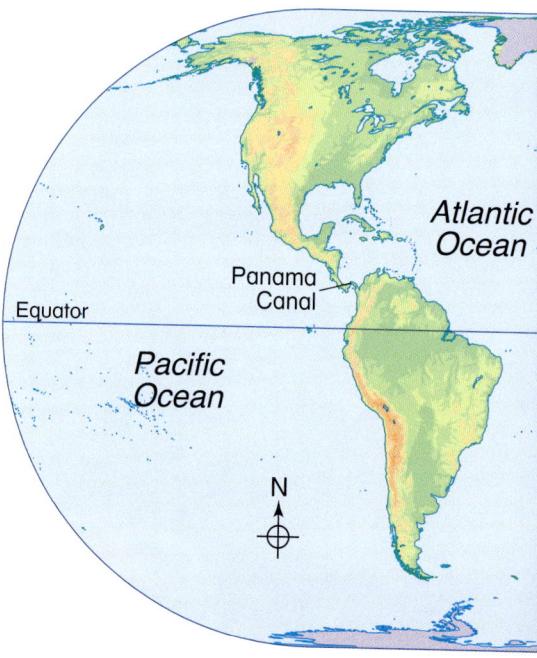

Discussion
- In what ways are oceans threatened?
- Why do threats to the ocean environment matter?
- How can the oceans be protected?

Mapwork
Draw a map of the Arctic Ocean. Add notes about the way it is threatened or changing.

Overfishing
Many people like to eat fish but supplies are getting harder to find. This is because modern fishing boats drag long nets through the water catching everything in their path. In the Atlantic Ocean, stocks of herring and cod are very low. Around Antarctica so many whales have been killed that they are in danger of extinction.

▶ Sorting fish from a factory ship.

Unit 1 Seas and oceans

Shipping

Most of the goods that are moved around the world are carried by ships. Tankers and bulk carriers are loaded with oil, coal, iron ore and other heavy cargoes. Some of the most important shipping routes are around the coast of Europe, Southeast Asia and the US. If ships collide or run aground it can cause terrible pollution.

▶ A container ship in Hong Kong harbour.

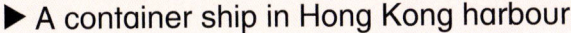

Pollution

The oceans are vital to the health of the planet. However, scientists are concerned that they are being used as rubbish dumps. Plastic and garbage are building up where currents are weak. Sea water is getting more acidic. In the Arctic Ocean old nuclear submarines have been left to rot on the sea floor.

▼ Clearing up beach plastic.

Investigation

Design a poster about threats to the ocean environment.

Unit 1 — Seas and oceans

Lesson 3: Learning about seas

What is a sea?

Around the edge of the oceans, there are places where the water is quite shallow. These are known as seas. The beaches and shore are often popular with visitors who come to enjoy the seaside.

Like oceans, seas are important habitats for fish and other animals. In some parts of the world, oil, gas and minerals have been found in rocks under the seabed. All these resources are very valuable.

▼ A coral reef in the Caribbean Sea.

▼ A North Sea wind farm provides clean energy.

Discussion
- How is a sea different from an ocean?
- Why are seas so useful?
- What impact have people had on the North Sea?

Mapwork
Using an atlas, make a list of seas around the world.

▼ A seaside resort in Greece.

Key words

fish stocks
oil platform
resort
resource
shore
wind farm

Unit 1 **Seas and oceans**

The North Sea

The North Sea is one of the busiest seas in the world. Thousands of ships cross it every day and 80 million people live around its shores. The North Sea is famous for oil platforms and wind farms. It is also home to many creatures, including seals, sea birds and fish.

Data bank
- On average the North Sea is only around 100 metres deep.
- There are around 170 oil and gas platforms and 40 wind farms in the North Sea.
- North Sea fish stocks are threatened by pollution and overfishing.
- Large numbers of sea birds have died due to bird flu.

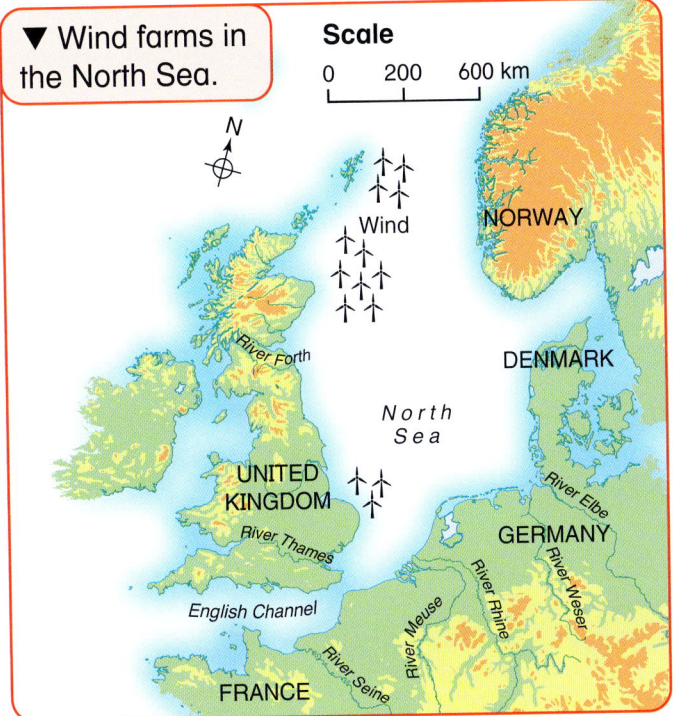

▼ Wind farms in the North Sea.

▲ Most parts of the North Sea are very shallow.

Mapwork
Make your own map of wind farms in the North Sea.

Summary
In this unit you have learnt about:
- what lies beneath the surface of the ocean
- how oceans are threatened
- conserving seas and oceans.

Unit 2 — Wearing away the land

Lesson 1: Rivers in action

How do rivers shape the land?

Key words
channel
deposition
erosion
reservoir
river bank
transportation
water cycle

▲ This canoeist is being carried along by the force of the water. All he has to do is steer clear of the rocks!

Data bank
- The way that water moves round the world is called the water cycle.
- The amount of water in the world always stays the same – no new water will ever be formed.

Unit 2   Wearing away the land

▼ **Erosion:** Rivers cut into the land creating valleys with steep sides.

As streams and rivers flow downhill, they remove tiny pieces of rock on the river bed. They also eat into the earth banks on either side of the channel. The tiny particles of rock and earth bounce and scrape along the river bed, wearing it away even more. This shapes and moulds the land over thousands and thousands of years.

A lot of the material which is carried along by the water is dropped somewhere else. Some of it slowly builds up into banks of sand, mud and gravel in the middle of the river. In other places, the material is dropped in lakes and reservoirs. Over long periods of time, these fill up and turn into dry land.

▲ **Transportation:** Flood waters are so powerful they can carry rocks, boulders and whole trees downstream.

▼ **Deposition:** Rivers drop gravel and mud.

Discussion
- What is the water cycle?
- How do rivers wear away the land?
- How do rivers build up the land?

Investigation
Make a drawing and write a sentence about three key words on this page in your geography notebook.

Unit 2   Wearing away the land

Lesson 2: Preventing flood damage

How can we control rivers?

The Mississippi River is over 3000 km long. It flows southwards across the United States, draining half the country. The Mississippi is an important route for shipping.

In the past, the Mississippi was very shallow and used to flood after heavy rain. The river was also constantly changing its channel as it meandered towards the sea.

Early in the 20th century, a team of river engineers made the channel deeper to help shipping. They also built banks to protect nearby farms, towns and factories from floods.

▲ Cargo boats on the Mississippi.

Key
- Over 1000 metres
- 200–1000 metres
- 0–200 metres

10

Unit 2 Wearing away the land

Dykes

Dykes along one side of the river force the water to cut a deeper channel on the opposite side.

Levees

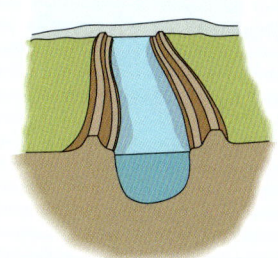

Huge earth and clay banks called levees hold back the flood water.

Cut-offs

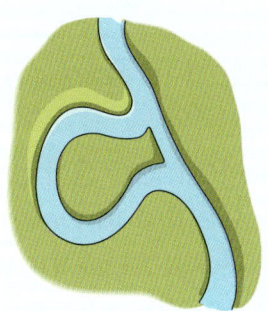

New channels cut off some of the meanders so the water can flow faster.

Boxes

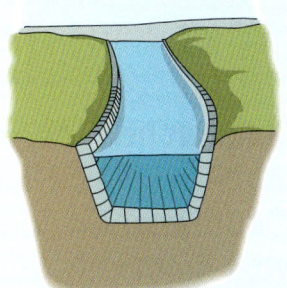

The sides and bottom of the channel are lined with concrete boxes to make them stronger.

Today there is a system of dykes, levees and dams along the Mississippi. The channel has been straightened and lined with concrete boxes. However, serious floods still happen. Some people question whether it will ever be possible to tame the Mississippi. They think the river should be allowed to flood as it did in the past and that the levees and boxes only trap the water and make things worse.

Climate change
Climate change is causing rivers to flood. Planting trees upstream and allowing rivers to meander are two ways of working with nature to reduce the problem.

▼ Mississippi flooding in Illinois.

Key words

channel
dyke
flood
levee
meander

Unit 2 Wearing away the land

Lesson 3: Finding out about rivers

What data is needed to find out about a river?

Children from Avenue Primary School went to a Field Study Centre in Devon (Southwest England). As part of their work on the environment, they found out about the nearby River Dart. The children tried to answers these questions.

River survey

1. In which compass direction does the river flow?
2. How wide is the river at three different places?
3. Is the river bed flat between the banks?
4. Are there any particles being carried along by the water?
5. Does the water flow at the same speed in the middle of the river as it does at the edges?
6. What plants grow on the river bank?
7. Is there any evidence of fish or animal life?
8. Are there any clues that the water level changes?

Discussion
- Why did the children go to the Field Study Centre?
- Which question in the list do you think is the most important?

▼ The children used this equipment when they studied the River Dart.

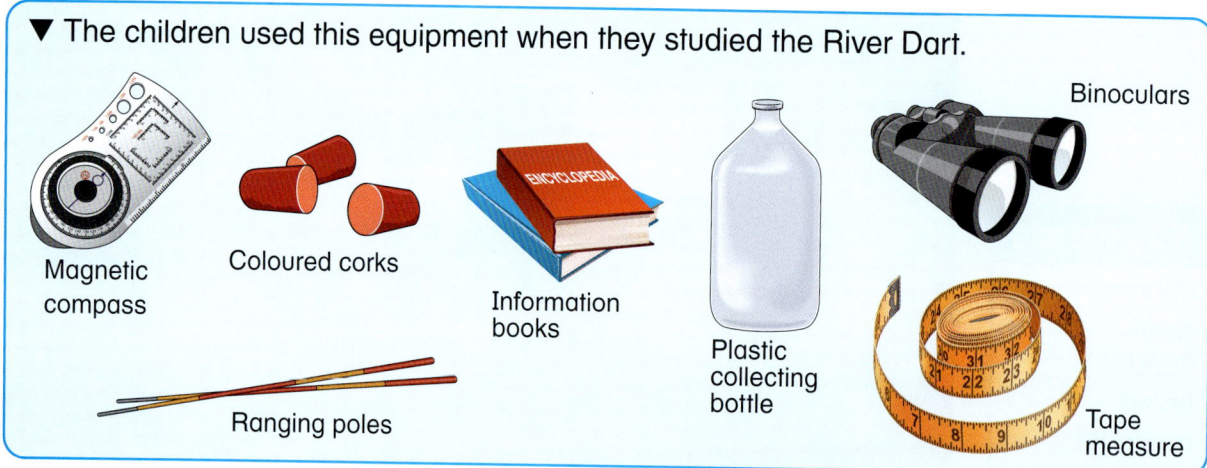

Unit 2 Wearing away the land

▲ Taking river measurements with a tape.

Our River Study

We spent Tuesday at the river. We had to measure carefully and make observations using a compass. We found that the river flowed from north to south. In places it was three metres wide. The ranging pole showed the water was half a metre deep. It was flowing quite fast. We could see water weed and animal burrows along the bank.

When the children returned to the Study Centre, they recorded their results on a spreadsheet. They also looked at local maps to trace the route of the river from source to mouth. Then the children also found out more about the plants and animals which they had seen on the river bank.

Mapwork
Using a local map, locate and name the nearest streams and rivers to you.

Key words	
environment	particle
magnetic compass	ranging pole

Investigation
Decide which piece of equipment the children would need to answer each survey question.

Summary
In this unit you have learnt:
- that rivers are a major influence on the landscape
- how people try to control rivers
- how to study a river.

Unit 3: The seasons

Key words: cycle, pattern, season, temperature, crop

Lesson 1: Changing seasons

What are the seasons?

Over the year there is a pattern to the weather depending on the season. In some countries, there are four seasons: in winter, the weather is often cold and the days are dark and short. In summer, the weather is much warmer and the days are long and bright. Spring is the time when plants begin to grow and birds build their nests. Fruit and other crops are harvested in the autumn.

In many countries, there are fewer separate seasons, such as Jamaica, which you will learn about in Unit 9. Seasons are different based on where you are in the world. The changing seasons give a pattern to our lives.

Discussion
- Which season is shown in each of the photographs on page 15?
- Which is the coldest and warmest month in the UK according to the temperature chart?
- How do the seasons affect people and plants?

Investigation
Cut out a circle of card to make a seasons dial. Add drawings and notes for each of the four seasons.

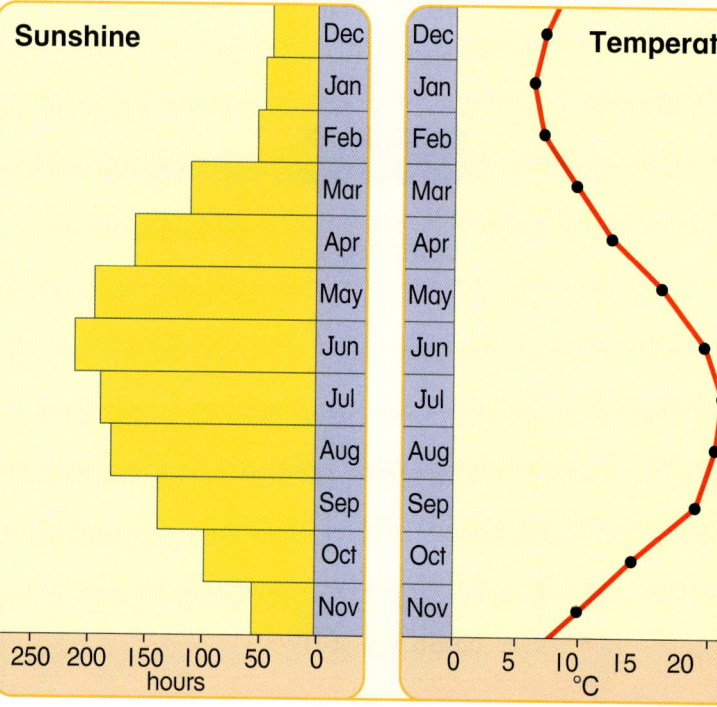

▼ Changes in sunshine and temperature in the UK affect the life cycle of animals, such as frogs, and plants.

Frog life cycle:
- Frogs hibernate
- Frogs lay spawn
- Tadpoles turn into young frogs
- Frogs slowly grow bigger

Unit 3 The seasons

Unit 3 The seasons

Lesson 2: Seasons worldwide

Do all places have the same seasons?

Key words
climate
Mediterranean
monsoon
season
tropical

In the United Kingdom there are four seasons which each last three months. Some other parts of the world have a different pattern of seasons. This affects how people live and the crops they can grow.

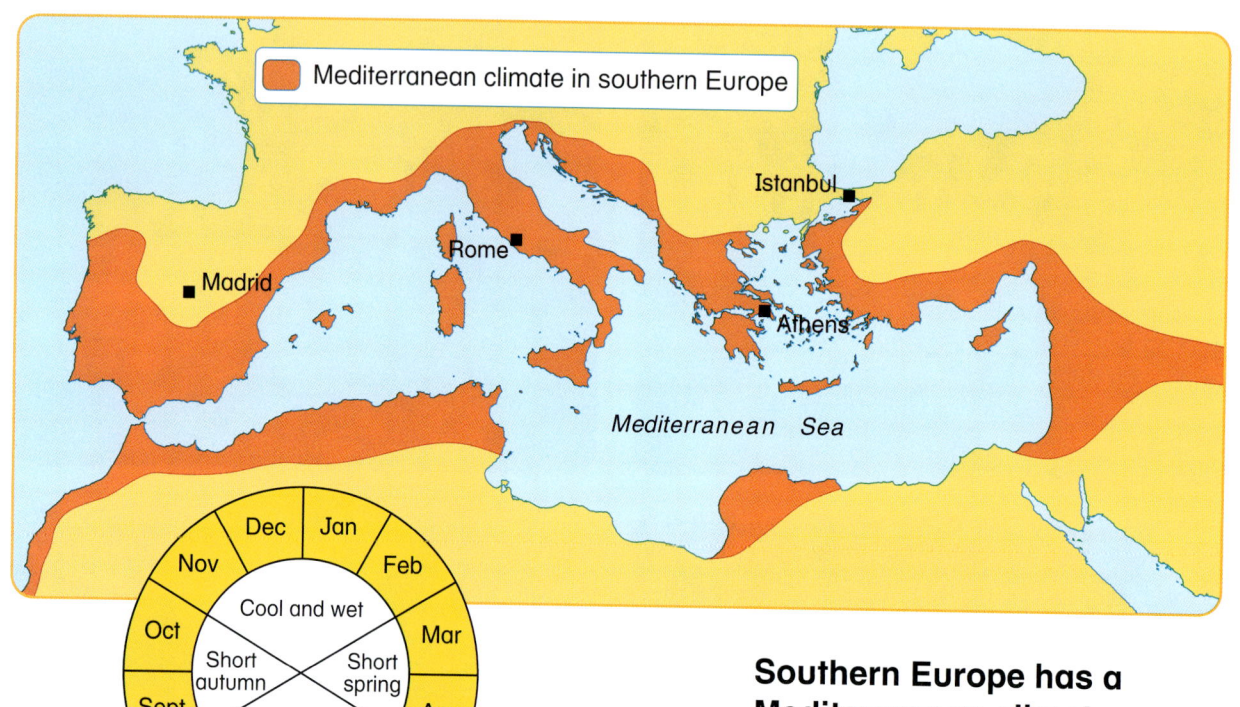

Southern Europe has a Mediterranean climate

Summer in the Mediterranean.

"Summer is a time of scorching heat. The countryside is alive with the humming of insects and the crackling of dry grass. In the fields the crops are ready to harvest.

During the day the whole land is flooded with light. The glare of the sun is thrown back from the white rocks. The streams dry up and only the tough plants can survive in the heat."

(Adapted from *Memed, My Hawk* by Yashar Kemal)

▲ Olive trees grow well in the Mediterranean.

Unit 3 The seasons

Discussion
- How long do seasons last in the UK?
- How does the Mediterranean climate differ from the UK?
- What is the effect of the monsoon in Asia?

Mapwork
Using an atlas, make a list of other parts of the world that have Mediterranean and monsoon (tropical) climates.

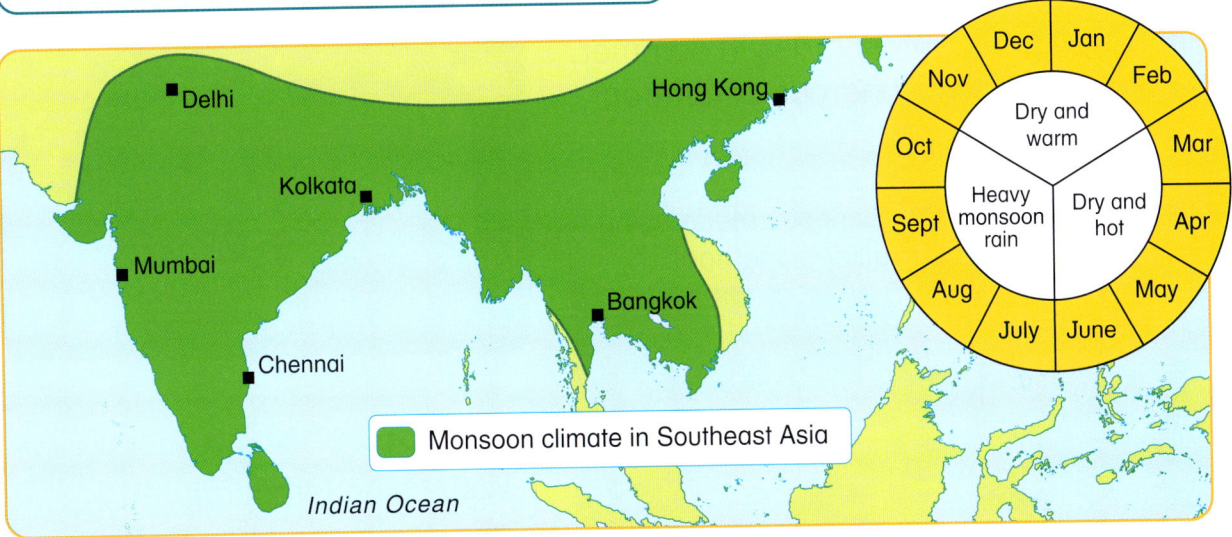

Southeast Asia has a monsoon (tropical) climate

The monsoon rains arrive.

"A hot wind blew through our bungalow day and night from the huge open plain. Then the clouds began to bank up and bank up and there was an unbearable feeling of pressure.

The rains came down with terrific force, such as you hardly ever see in Europe. This would probably go on for two or three days and the whole area round the houses turned green. An extraordinary life burst out."

(Charles Allen)

▲ Monsoon rains are vital for crops such as rice. One thousand million people depend on rice grown during the monsoon for their food.

Investigation
Write a description of the weather for one of the seasons in your country. Ask a friend to guess which season.

Unit 3 The seasons

Lesson 3: Seasonal influences

How are farmers affected by the seasons?

Key words
dairy
grazing
harbour
high altitude
pastures
resort

The mountains of Switzerland are famous for making cheese. Farmers keep cows, sheep and goats which graze on the mountain grass in summer. Dairy farms make hundreds of different types of cheese.

▼ Cows grazing in the Swiss Alps.

▼ Cheeses made using traditional methods.

Climate change
With rising temperatures and changes to weather, less grass is growing in the fields. This is affecting Swiss cheese farmers.

	Spring	Summer	Autumn	Winter
Cows	Cows taken to high altitude pastures to graze.	Cows spend the summer in the high pastures.	Cows continue to graze in high pastures in mild weather.	Cows return to the valley farms for protection, where they feed on hay.
Milk	Milk production is high during the grazing months.	Milk is made into cheese.	Milk production is high and cheese continues to be made.	Milk production reduces.
Haymaking	Grass grows in the fields.	Grass grows in fields and is harvested for hay.	Hay is harvested and stored in barns.	Hay is used to feed cows in winter.

Mapwork
What season is it for you now? Plan a short 'seasonal walk'. You might find flowers, leaves, berries, puddles, for example. Note where you found these seasonal clues on a simple sketch map.

Investigation
Make a seasons chart for your own area. Write down three things for each season that you might notice or do.

Unit 3 The seasons

How are seaside resorts affected by the seasons?

Whitby is a town on the coast of Yorkshire in England. It has a busy fishing harbour and a long sandy beach. Many people go to Whitby for their holidays. Some people visit the old abbey.

▼ Whitby harbour.

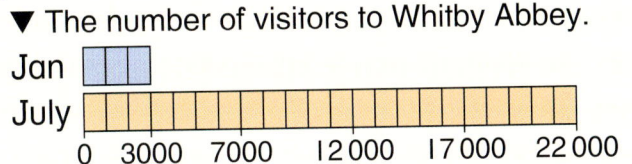
▼ The number of visitors to Whitby Abbey.
Jan
July
0   3000   7000   12 000   17 000   22 000

Summer
In summer Whitby is crowded with visitors who come from all parts of the UK and abroad.

Winter
In winter Whitby is much quieter, particularly when the wind brings rain and snow from the North Sea.

- Beach busy with tourists.
- Boat trips.
- Steam railway.
- Museum open every day.

- Beach empty.
- Sea too rough for boat trips.
- Railway line repaired.
- Museum only open at weekends.

Discussion
- Which is your favourite season and why?
- How does Whitby change between summer and winter?
- Is the weather this week typical for the season?

Summary
In this unit you have learnt:
- how the seasons are different
- about seasons around the world
- how people are affected by the seasons.

Unit 4 Cities

Lesson 1: Describing cities

What are cities like?

Key words
city centre
settlement
skyscraper
suburbs
vandalism

Climate change

Well-designed cities can help to combat climate change. When lots of people live close together (a) it is easier to collect and recycle waste (b) their homes use less energy and (c) people can walk or cycle to places. Which of these things do you think matters most?

▲ Doha, Qatar.

Unit 4 Cities

Cities are the largest of all settlements. They are busy, crowded places with hundreds of thousands of homes.

During the day, people come to the city centre to buy things in shops and work in office blocks. As night falls, people leave work and go to restaurants, theatres, cinemas and clubs. The streets are full of bright lights from the shops and advertisements. Away from the centre, it is quieter. The suburbs spread out into the countryside. There is more space there for houses and gardens, shops and parks.

Cities are connected to other places by high-speed trains, motorways and airports. The roads and underground trains are often crowded. Noise and fumes can be a problem. Some areas suffer from vandalism.

Data bank
- The Tacoma Building in Chicago, US (1889) was one of the world's first skyscrapers.
- Hong Kong has more skyscrapers than any other city in the world.

Investigation
Draw a picture of yourself. Add a speech bubble saying whether or not you have been to a city. What do you think about city life?

Discussion
- Why do people want to visit city centres?
- Why are there suburbs around cities?
- Why do you think cities are getting larger?

Unit 4 Cities

Lesson 2: World cities

How are cities changing?

Around the world, many cities are getting bigger as people move in from the countryside. Some people are attracted to cities because they think they will find a better life there. Others are forced to move out of their villages by floods, drought, war and famine.

City	Millions of people	
	2000	2020
Tokyo, Japan	34	37
Delhi, India	16	30
Shanghai, China	14	27
Mumbai, India	16	20
Mexico City, Mexico	18	22
Cairo, Egypt	14	21
São Paulo, Brazil	17	22
Beijing, China	10	20
Dhaka, Bangladesh	10	21
Osaka, Japan	19	19

Key words		
city	countryside	population

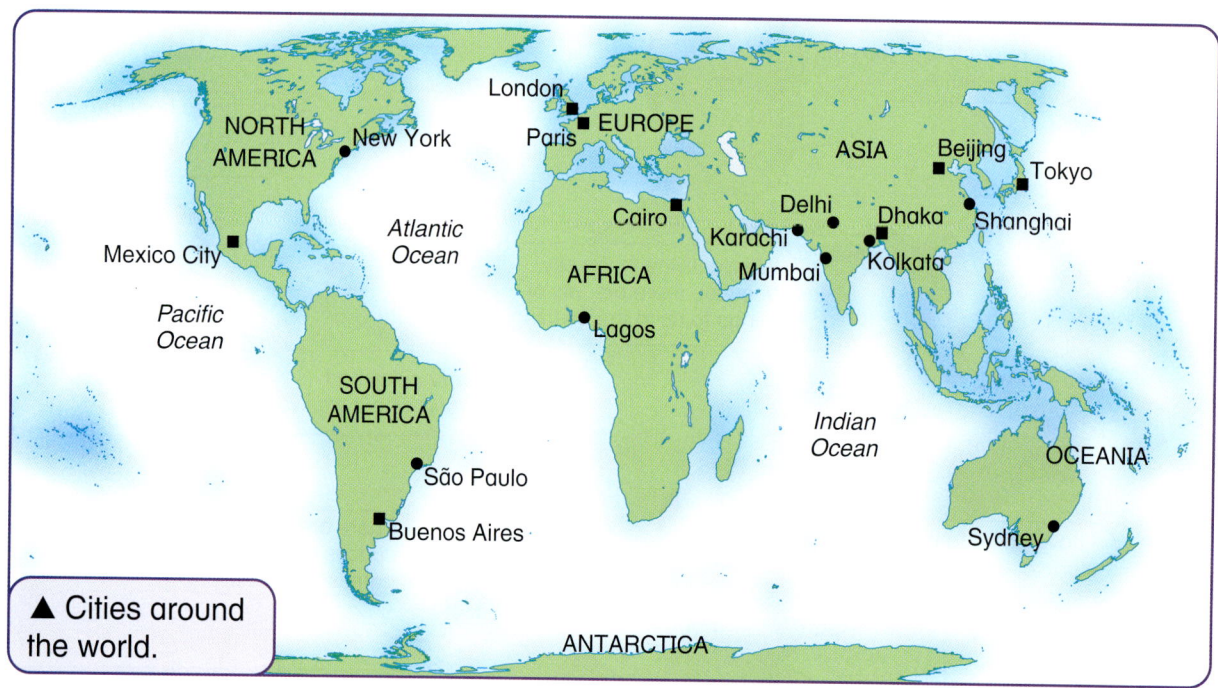

▲ Cities around the world.

Discussion
- Which cities do you think will be the three largest cities in the world in the next five years?
- Why do cities attract people?
- What is special about the city nearest to you?

Investigation
Using the data from the world cities table above, draw a bar chart of the world's city populations.

Unit 4 Cities

The story of New York

New York is on the east coast of the United States. It began as a trading post set up by Dutch sailors nearly 400 years ago. Now it has a population of over 20 million people speaking 800 languages and is an important centre for world trade and finance.

New York is famous for the Statue of Liberty and the Empire State Building. Central Park is a popular meeting place. The museums of modern art and natural history attract many visitors. In the evening there are shows to see in the 40 theatres along Broadway.

New York enjoys sunshine on most days in the year but January can be very cold and sudden deep snow is not unusual.

Mapwork
Find six more world cities then label them on a world map.

Data bank
- The Empire State Building has 102 floors and 6500 windows.
- In New York, people spend over one billion dollars on theatre tickets each year.
- There are 70 000 helicopter flights over New York each year.

▲ New York City.

Unit 4 Cities

Lesson 3: The story of London

Key words
crossing point
Thames
underground

How has London grown and changed?

London was built at a crossing point on the River Thames. Low hills on the north bank of the Thames provided a dry site above the marshes. Ships could sail up the river carrying goods from other parts of Europe.

▼ Old Roman wall.

0 CE

200

The Romans built a small town with a temple, forum and harbour. They also built a wall around the town to protect it.

400

600

By Norman times, London was the capital city of England. William the Conqueror had the Tower of London built for defence.

800

▲ The Tower of London.

1000

1200

In 1666, the Great Fire destroyed large parts of the city. St Paul's Cathedral and many other fine new buildings were put up.

1400

▼ St Paul's Cathedral.

1600

1800

By the 19th century, London had become a large industrial centre and the capital of a worldwide empire.

2000

▲ Tower Bridge.

Unit 4   Cities

▼ A tourist map of landmarks in central London.

Investigation
Find out one thing about each of the cities marked on the map of the UK.

Data bank
- London was the first city in the world to have underground trains.
- There are over 30 bridges spanning the Thames in London.
- Four out of ten of the people who live in London were born abroad.

Mapwork
Draw a map of a walk through a city linking four different landmarks.

Climate change
There are eight million trees and lots of parks and gardens in London. It became the world's first National Park City in 2019 because parks are good for people and the environment.

Summary
In this unit you have learnt:
- how cities are different from other places
- how cities are changing
- how to describe a city.

25

Unit 5   Jobs

Lesson 1: Making things

Where are things made?

Most of the things that we eat, wear and use each day are made in factories. Factories can be very small places with one or two workers or very large places with thousands of workers.

Factories use machines to make things quickly and cheaply. Machines can make large numbers of things in exactly the same way each time.

Key words	
dough	storage
ingredients	waste
input-output	workshop
raw materials	

All factories need:

- workshops, offices and storage space
- an entrance, delivery area and car park
- power for heating and machines
- computers to process information
- a way to recycle or get rid of waste.

Discussion
- What do factories do?
- What things around you might have been made in a factory?
- What is the difference between a factory and a school? Is anything the same?

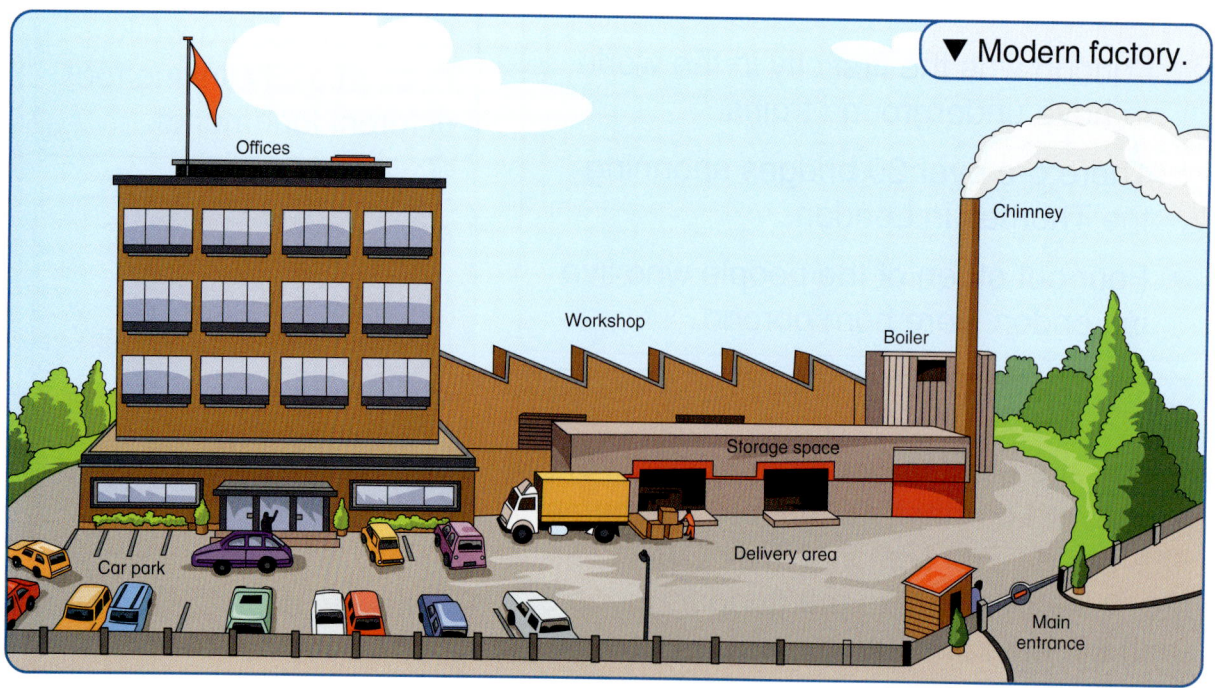

▼ Modern factory.

Unit 5   Jobs

How do factories work?

Before it can make anything, a factory needs raw materials. These materials are either dug out of the earth or produced by farmers and fishers. Factories turn these raw materials into the goods we see in shops.

Investigation
Draw Input-output diagrams for making a pencil and a carton of apple juice.

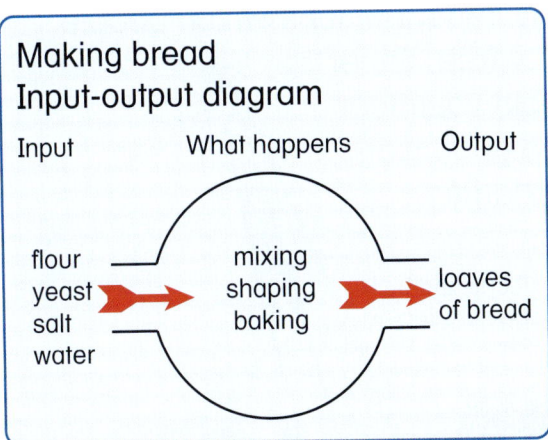

Making bread
Input-output diagram

Input — What happens — Output

flour, yeast, salt, water → mixing, shaping, baking → loaves of bread

▼ Special clothes keep bakers clean.

▲ Bread is made from flour, yeast, salt and water.

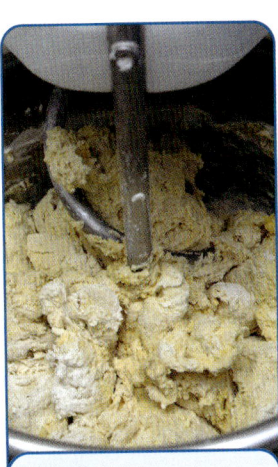
▲ The ingredients are mixed to make dough.

▲ The dough is shaped into loaves and baked in the oven.

▲ The bread is sliced, and put into crates.

INPUT → WHAT HAPPENS → WHAT HAPPENS → OUTPUT

Unit 5 Jobs

Lesson 2: Different jobs

How do people earn a living?

Key words
bulk carrier
harbour
lifeboat
skill

Most people who do a job are paid for what they do. The amount they are paid depends on their skills, their qualifications and the number of hours they work.

There are different types of jobs. Some people work in offices, collecting information, using computers, or making plans. Others work out of doors where they drive machines or farm the land.

In this lesson, you can find out more about jobs that are done in a harbour.

Most people are paid for what they do, but some are not. Caring for children or older people at home are really important jobs which people do for free. Other people volunteer to do good work for charity.

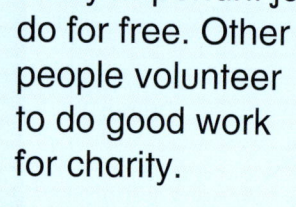

Joan Lovell works in a factory. She cleans and boils crabs, lobsters and other shellfish so they are ready to be packed to send to the shops.

Bill Shaw is a fisher.

Unit 5 Jobs

Discussion
- What is a harbour?
- Which job is the most important?
- Which job do you think is most interesting?

Investigation
What single special skill does each job require? Show your answers in a table.

Mapwork
Make a plan of the harbour from the clues in the picture.

Jack Perez is a crane driver. He unloads cargoes from the boats.

Carla Gulati is the harbour master. She gives permission for boats to come in and out of the harbour.

Steven Bell is the coxswain (captain) of the lifeboat.

Winston Hayes is an engineer. He checks and maintains all the machinery in the harbour.

Susan Hoff drives a bulk carrier. It carries gravel chippings from the harbour to local road works.

Maria Bruni keeps records and accounts in the harbour master's office.

29

Unit 5 Jobs

Lesson 3: Types of work

What are the different types of work?

Key words
primary
retirement
secondary
tertiary

Jobs can be put into three main groups:

1. **Primary activities** involve collecting and harvesting natural resources.

2. **Secondary activities** involve making things from natural resources.

3. **Tertiary activities** involve providing services for people to use.

In the past most people around the world worked in primary activities. Now tertiary activities are more important. Technology and artificial intelligence is also changing the way people work.

▶ **Primary activities**
farming
fishing
forestry
mining

▶ **Secondary activities**
making clothes
putting up buildings
building roads
making chemicals

Data bank

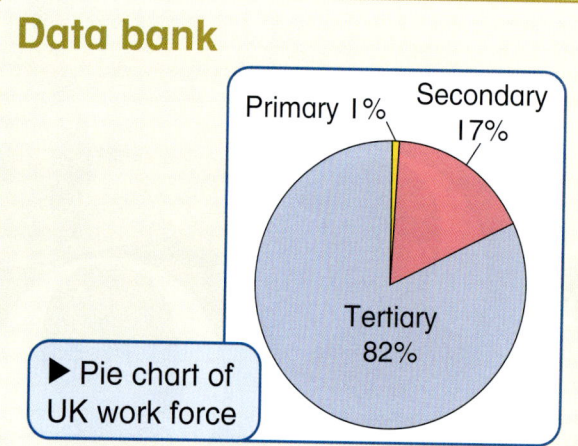

▶ Pie chart of UK work force

Primary 1%
Secondary 17%
Tertiary 82%

- In the UK women are still paid less than men for the same work.
- Some people are out of work and cannot find jobs.
- The retirement age has increased from 65 to 68 or 70 years.

◀ **Tertiary activities**
shop work
office work
driving vehicles
caring for people
entertainment

Discussion
- What types of job do adults do in your school?
- Which type of work do you think is hardest?
- Why are fewer people needed in primary activities nowadays?

Unit 5 Jobs

In one school children used a local newspaper to make a survey of jobs in their area. They copied some of the advertisements and put them into a scrapbook using separate pages for primary, secondary and tertiary work.

Investigation
Make a scrapbook of your own using the advertisements from this page, together with ones from a local paper.

VAN DRIVER
required by international food company.
We are looking for a committed and reliable person for this permanent job.

Chemist
Local company seeks bright young scientist for laboratory work. Duties include quality control and testing.

St Stephen's Junior School
Secretary
Required for this busy, popular friendly school. 35 hours per week. Office experience and knowledge of Word processing and spreadsheets an advantage.

WE NEED YOU
STRAWBERRY PICKERS
(from mid May)
Good rates of pay. Transport available.

City centre restaurant requires young chef

Printer
Full time 9.00am to 5.30pm. Must be able to work on their own. Graphic design knowledge an advantage.

ASSEMBLY LINE WORKERS
Modern company has four full time positions. Light work, training provided.

Leisure Homes
require a Swimming Pool Attendant. Qualification required.

Sales Manager
needed for expanding local business. Must be willing to travel.

Coach Driver needed for summer season. Licence required.

Office Person
Busy office. Good communication skills essential.

Nursing Home requires a Chief Nurse
This is a new post which would be ideal for an experienced nurse.

Mapwork
Show where different jobs are done on a simple outline map of your school. You could add labels or use symbols or a colour code.

Summary
In this unit you have learnt:
- what factories do
- how jobs are linked together
- about how jobs can be grouped.

Unit 6　Pollution

Lesson 1: Damaging the environment

What causes pollution?

When fumes, noise and waste cause damage to the environment it is called pollution. Some natural events cause pollution. For example, if a volcano erupts it pollutes the air with large quantities of dust and poisonous gas. However, many pollution problems are caused by people.

Key words	
chemicals	pollution
nuclear waste	volcano

Discussion
- Where does pollution come from?
- Which types of pollution do you think are most serious?
- How does pollution affect people and animals?

▼ Every year eight million tonnes of plastic is dumped into the sea. This washes up on beaches all over the world.

Unit 6 Pollution

How do we cause pollution?

Many of the things I buy are made of plastic.

Some of the factories which make cheap goods put fumes into the air.

▲ Air pollution

I need to wash and keep clean.

Wastewater has to be purified before it is put back into seas and rivers.

▲ Water pollution

I like eating crisps.

We all need to eat but the rubbish from food packets has to go somewhere.

▲ Land pollution

How long does pollution last?

Pollution does not last forever. Fruit and vegetable peelings will rot in a few weeks. Paper and cardboard disappear in about a year. Metal lasts for many years because it rusts slowly.

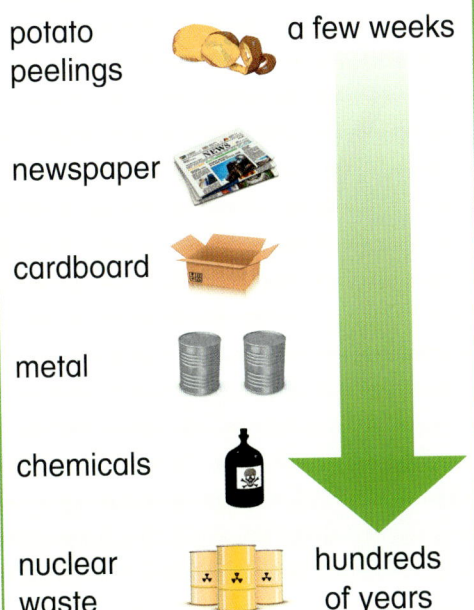

potato peelings — a few weeks

newspaper

cardboard

metal

chemicals

nuclear waste — hundreds of years

Mapwork

Make a map to show clean and dirty areas in your school building or local surroundings.

Investigation

Choose six objects in your home or classroom. Make a chart to show how long each one might take to decay.

Data bank

- On average, each person in the UK throws away their own body weight in rubbish every seven weeks.
- A plastic bag can take up to 100 years to decompose.
- Nuclear waste can take between 10 000 and a million years to decay.

Unit 6 | Pollution

Lesson 2: 'Green living'

How can we reduce pollution?

Key words	
carbon emissions	renewable resources
dam	solar panels
recycle	turbine
	wind farm

We can all help to reduce pollution. We could buy fewer things which would save resources. We could also make sure that the things we no longer need are recycled.

Governments have passed laws to protect the environment and reduce carbon emissions that change the climate. The trouble is that it takes a long time for people to change their habits.

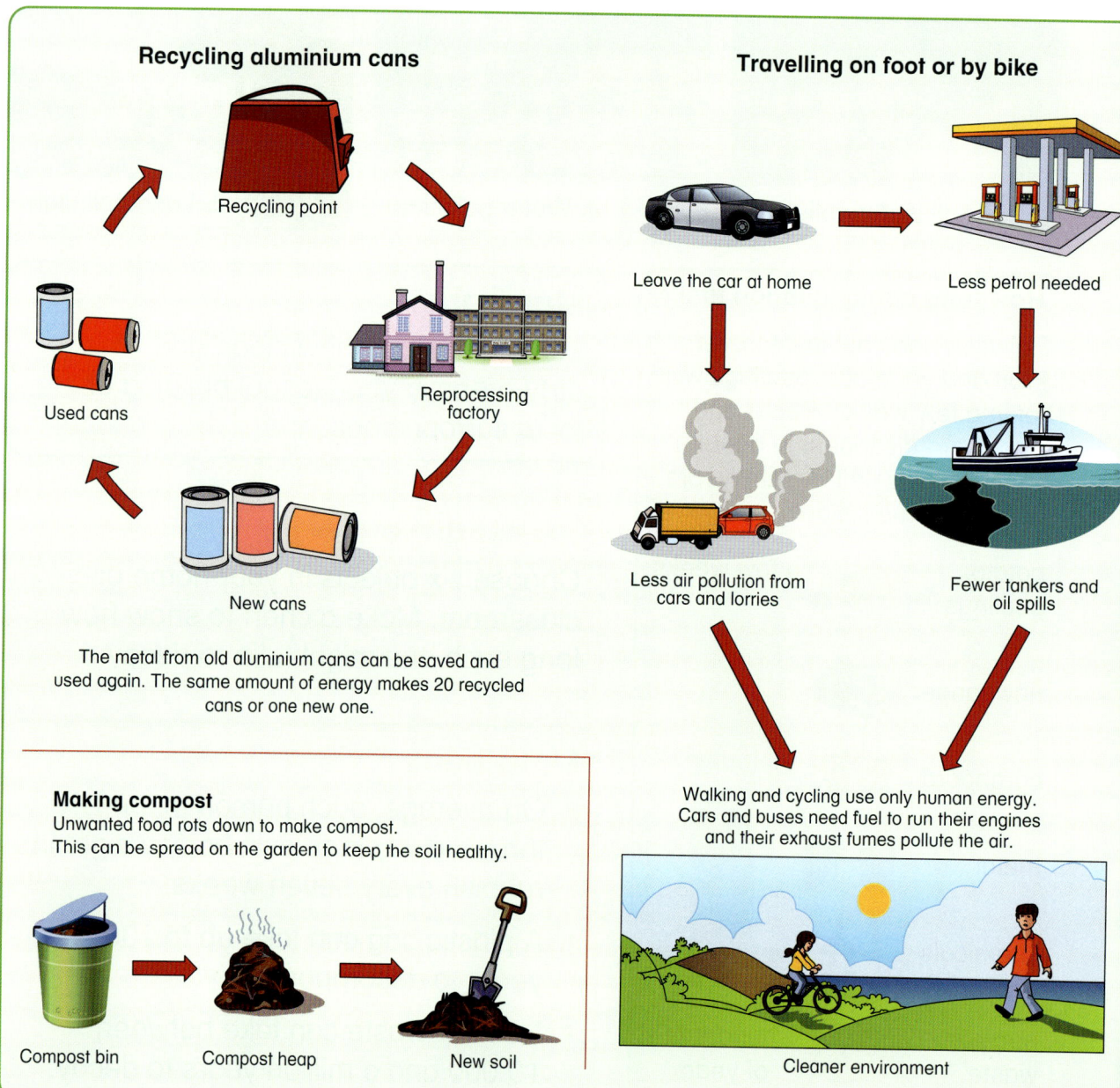

Recycling aluminium cans
Recycling point
Used cans
Reprocessing factory
New cans

The metal from old aluminium cans can be saved and used again. The same amount of energy makes 20 recycled cans or one new one.

Making compost
Unwanted food rots down to make compost. This can be spread on the garden to keep the soil healthy.

Compost bin — Compost heap — New soil

Travelling on foot or by bike
Leave the car at home
Less petrol needed
Less air pollution from cars and lorries
Fewer tankers and oil spills

Walking and cycling use only human energy. Cars and buses need fuel to run their engines and their exhaust fumes pollute the air.

Cleaner environment

34

Unit 6 Pollution

Discussion
- What does 'recycling' mean?
- Why does it take time to change people's habits?
- How could you save energy yourself?

Investigation
Devise a ten point 'Waste and Pollution' policy for your school.

Using 'green' goods

Washing powder and weed killer sometimes contain harmful chemicals. We can buy alternatives which do less damage to the environment.

Saving energy

If everyone switches off lights they do not need turns down the heating thermostat or air conditioning in their homes, it would save a lot of energy.

Renewable energy
There are ways to make electricity causing very little pollution. These use natural forces like the heat from the sun and the power of wind and water. They are called 'renewable' because, unlike coal, oil and gas, they will never run out.

▲ Solar panels

Dams
Dams can trap water from rivers and seawater. When the water is released it is used to turn turbines.

Solar panels
Solar panels catch the energy from the sun and turn it into electricity.

Wind farms
In some exposed places, like marshes and the tops of hills, there are groups of wind turbines. When the wind blows, it turns the blades.

Climate change
There are three words beginning with 'r' which are key to 'green living': **r**educe, **r**euse, **r**ecycle. Think about each of these words in turn and the actions that might go with them.

Unit 6 Pollution

Lesson 3: Exploring clean energy

Key words
energy
fumes
power station

> Can old power stations make clean energy?

Drax power station is the largest power station in the UK. It has six boilers, twelve cooling towers and a very high chimney. It generates electricity by burning fuel to make steam that drives turbines.

In the past Drax was powered by coal. This caused a huge amount of air pollution and contributed to global warming. Drax has now stopped using coal and burns wood pellets instead.

Drax still has problems. The wood that it uses comes from Canada. Cutting down the trees, turning the wood into pellets and bringing it to the UK damages the environment. Some people worry that the trees will not be replanted and that burning the wood creates a lot of fumes.

Discussion
- What is a power station?
- Is it better to burn wood pellets rather than coal?

▼ Drax power station has 12 cooling towers.

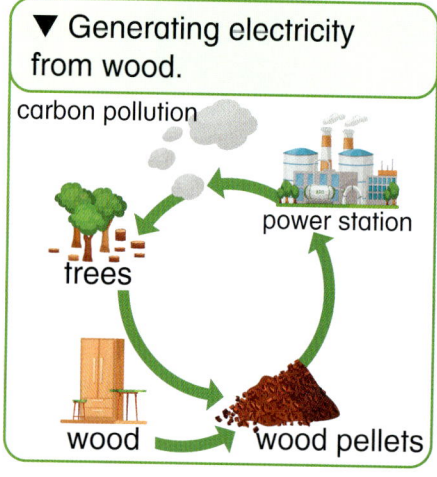
▼ Generating electricity from wood.

Data bank
- Drax used to burn around a million tonnes of coal a year.
- It now uses 6.5 million tonnes of wood pellets a year.
- Drax generates up to 15% of renewable energy in the UK.
- There are now plans to capture the fumes from burning the wood pellets.

Unit 6 Pollution

A local investigation

At one primary school, Class 5 decided to investigate local pollution problems. They recorded each problem with a circle on a survey sheet like the one opposite. They then added up the numbers they had circled. The score told the children how badly their local environment suffered from pollution. The children also took photographs and made a map for a class display.

Pollution survey

Problem	No Problem	Some	A lot
Traffic exhaust	0	5	(10)
Factory fumes	0	5	(10)
Traffic noise	0	5	(10)
Aircraft noise	(0)	5	10
Polluted water	0	(5)	10
Litter	0	(5)	10
Overhead wires	0	(5)	10
Unpleasant smells	0	5	(10)
Factory noise	0	(5)	10
Vandalism/graffiti	(0)	5	10
Total			60

0 — No pollution   25 — Some pollution   50 or more — A lot of pollution

▲ Fumes from traffic.

▲ Noise from road works.

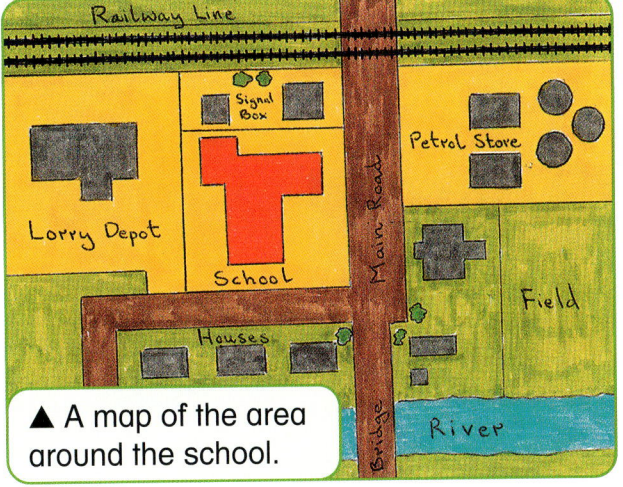
▲ A map of the area around the school.

Investigation
Make a similar survey of pollution problems around your school.

Mapwork
Draw a map or plan of the area you have studied in your pollution survey.

Summary
In this unit you have learnt:

- about different forms of pollution
- what people are doing to solve pollution problems
- how to study pollution.

Unit 7 Wales

Lesson 1: Mountains and valleys

What is Wales like?

Wales is a country of hills, mountains, moors and valleys. It lies to the west of England and is about 200 kilometres long and 100 kilometres wide.

In the past there used to be a lot of mines in Wales. Roof slates for houses came from mines in the mountains of North Wales. Coal for factories, ships and train engines came from South Wales. Today the mines have closed and new industries have taken their place.

Two out of every three people in Wales live in the south of the country. In central Wales, people live in small towns, scattered villages and farms. In the north, the main settlements are on the coast where tourism is important.

Key words

Anglesey
Cardiff
River Wye
Yr Wyddfa
   (Snowdon)
Swansea

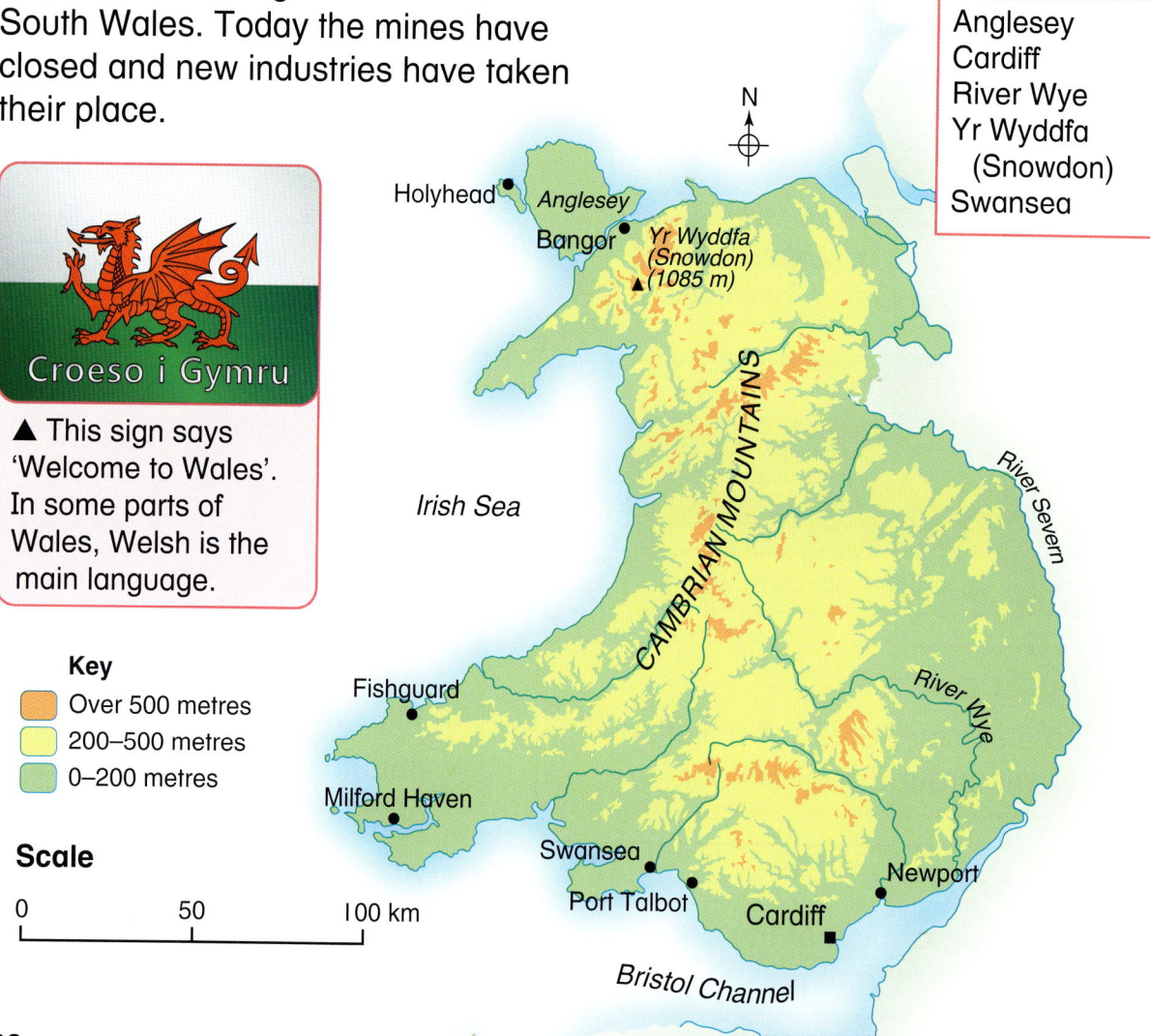

▲ This sign says 'Welcome to Wales'. In some parts of Wales, Welsh is the main language.

Key
- Over 500 metres
- 200–500 metres
- 0–200 metres

Scale
0   50   100 km

38

Unit 7 Wales

Discussion
- What are the main features of Wales?
- What used to come from Welsh mines?
- Why do you think most people in Wales live on the coast?

Rivers and landscape
- Yr Wyddfa (Snowdon) is the highest peak in Wales.
- The Wye is the longest river (209 km).
- Anglesey is an island off the northwest coast.

Transport
- The main airport is at Cardiff. The M4 motorway links Wales and England via the Severn Bridge.
- Ferries sail to Ireland from the ports of Fishguard and Holyhead.

Weather
- Most parts of Wales have a high rainfall.
- Strong winds affect the coast and other exposed places.
- On the highest mountains, snow lasts all winter.

◀ Hikers in Eryri (Snowdonia).

Settlement
- Cardiff is the capital city.
- Swansea and Newport are important ports on the Bristol Channel.
- Bangor is a seaside resort on the north coast.

▲ Cardiff Waterfront.

Work
- Most of the factories are in South Wales.
- In the mountains, most farmers keep sheep. There are dairy farms in the lowlands and valleys.
- Forestry is important in hilly areas in North and Central Wales.

Mapwork
Using the map on page 38, work out the distance from Cardiff to Bangor (a) by land (b) by sea.

Investigation
Using the information from pages 38 and 39 create a fact file for (a) North Wales (b) South Wales.

Unit 7 Wales

Lesson 2: The story of Blaenavon

How is Wales changing?

Key words
colliery
furnace
ironstone
lift shaft
heritage site

▲ The view over Blaenavon and Big Pit.

Blaenavon is an industrial town about 40 kilometres north of Cardiff, high in the mountains of South Wales. The name Blaenavon means 'the source of the river' in Welsh.

In 1789, during the Industrial Revolution, three furnaces were built at Blaenavon to make use of the local coal and ironstone. Within a few years the first coal mines had opened and the furnaces were producing thousands of tonnes of iron.

In 1852, in Victorian times, a railway was opened linking Blaenavon with Newport docks. The town grew larger as rows of stone cottages were built on the hillsides.

A large new colliery was also built. It was called Big Pit because it had a wide and deep lift shaft. The coal from Big Pit was excellent quality. It burnt with a great heat and left very little ash. This made it ideal for ships and railways engines all over the world.

▼ The ironworks in about 1800.

▼ Big Pit in the 1900s when demand for coal was at its highest.

Unit 7 Wales

As demand for coal decreased in the 20th century, many miners lost their jobs. When the iron and steel works closed, even more workers became unemployed. Big Pit was modernised to try to keep it open, but it finally closed down in 1980.

Today Blaenavon has a population of about 6000 people. There is a factory that makes parts for aeroplanes, and a family business that makes cheese. In 1983, Big Pit opened as a coal-mining museum. It was made into a World Heritage Site in 2000 and has won several important awards since then.

▲ A tourist map of Blaenavon.

Discussion
- How do you think turning Big Pit into a museum changed Blaenavon?
- What are the problems of burning coal?

Mapwork
List some of the differences between an Ordnance Survey map of Blaenavon and the map shown here.

Investigation
Make a timeline starting in 1789 to show the history of Blaenavon.

Unit 7 | Wales

Lesson 3: A visit to Big Pit

What was it like to be a coal miner?

Key words

coal face | pit head
fan house | shaft
lift shaft | ventilation

Glyn Davies used to work at Big Pit. He left school when he was 15. His job was to repair the steel ropes that pulled the lift up and down. Now he is a museum guide.

Glyn gives visitors a safety helmet with a cap lamp before taking them underground. He also gives each visitor a self-rescuer, just like the miners used to wear. Then everyone gets into the cage which quickly drops 90 metres to the bottom of the lift shaft.

It is very dark and musty in the mine. *"There are 40 kilometres of tunnels in Big Pit."* Glyn explains. *"They are held up with wooden props, bricks and steel arches. It was very important to get fresh air into the mine. A fan and ventilation doors helped to get rid of dangerous gases."*

Discussion
- How long did miners work in each shift?
- Why is it important to have museums like Big Pit?

▼ Visitors wearing their safety helmets.

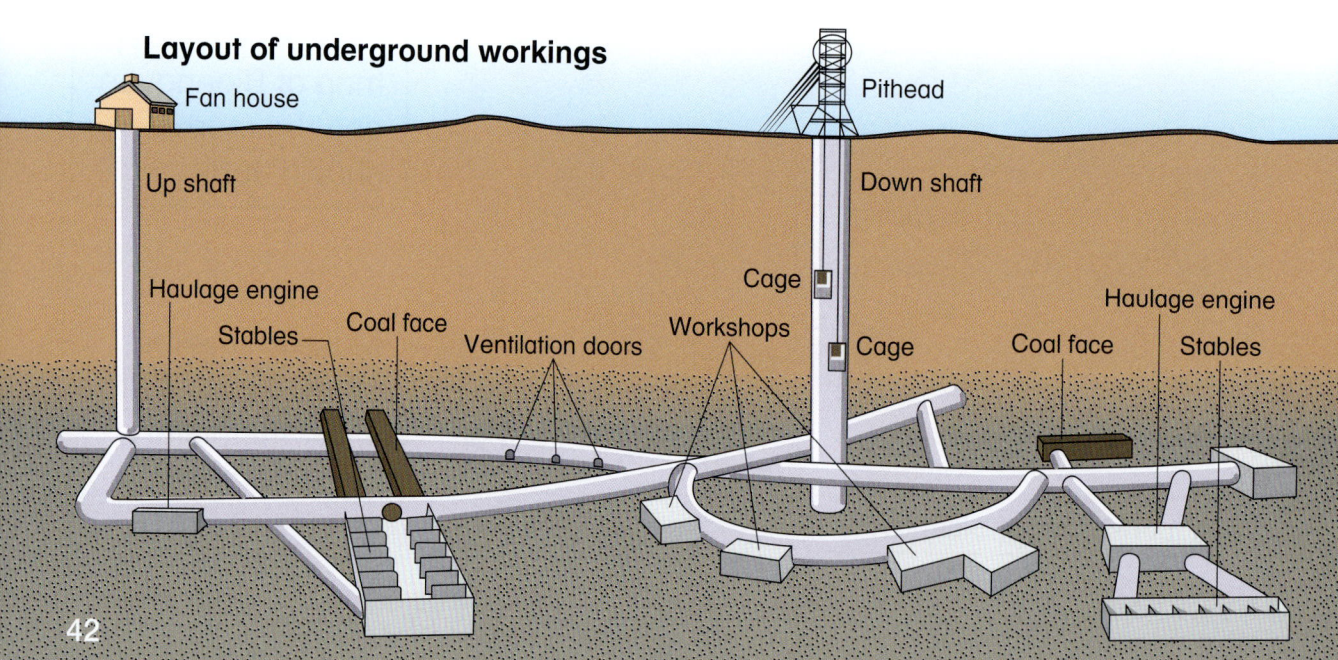
Layout of underground workings

Unit 7 Wales

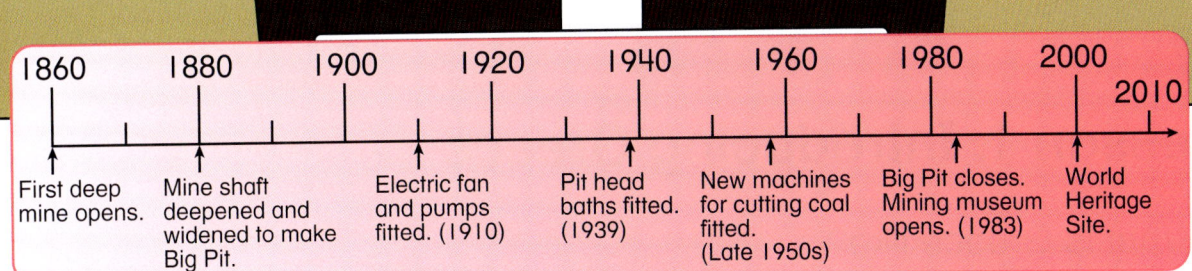

Glyn takes the visitors along the tunnels. He shows them the coal trucks and the rails they ran along. In some places, the ceiling is so low that everyone has to bend down. Water drips from the roof and the rocks are stained orange from the iron in the water. At one point, Glyn asks everyone to turn out their lamps. It is very dark. Then Glyn takes the visitors to the stables. In the past, there were over 70 ponies at Big Pit. They lived in the dark all their lives, pulling coal trucks.

Finally the visitors come to the coal face. Men worked in eight-hour shifts there. At night they cut away the coal. The morning shift loaded the trucks and the afternoon shift propped up the ceiling to make the new tunnel. It was a dangerous job and dust used to get into the miners' lungs, which made them ill.

"As far as I am concerned," Glyn says, *"the coal from Big Pit was the finest in the world. Also we had very few accidents. The problem was that the coal seams ran out so the mine had to close. It was tough work but I am proud to have been a miner."*

Data bank
- At one time over a thousand miners used to work at Big Pit.
- The coal face is around 100 metres below ground.
- Over three million visitors have visited Big Pit since it became a museum in 1983.

Investigation
Find out about other World Heritage sites.

Mapwork
Make a 'spoke chart' with your school in the centre. Show museums, parks, visitor centres and other attractions children in your class have visited at the end of each spoke.

Summary
In this unit you have learnt:
- what makes Wales different from other countries in the United Kingdom
- about the workings of a coal mine
- how some parts of Wales are changing.

Unit 8 Greece

Lesson 1: Introducing Greece

What is Greece like?

Three thousand years ago a great civilisation flourished in Greece. The ancient Greeks made beautiful temples, statues and theatres. They also made important discoveries in science and mathematics. People have admired and copied their ideas ever since.

Today, Greece is part of the European Union. In addition to the mainland there are hundreds of islands scattered across the Aegean Sea. These are popular with tourists who come to enjoy the summer sun.

Key words

Aegean Sea	European Union
Athens	Mediterranean climate
Crete	Pindus mountains

Mapwork

Using an atlas, work out how far it is from Athens to London and other European capitals.

Unit 8 Greece

Landscape

Much of Greece is covered by rocky mountains. These reach out into the sea in chains of islands.

▲ The Pindus Mountains.

Settlement

One in three people live in Athens, the capital city. Piraeus and Thessaloniki are the main ports.

Transport

Road and rail routes tend to follow the coast. Ferries link the islands with the mainland.

▶ The Corinth Canal links the Aegean and Ionian Seas.

Climate

Greece has a Mediterranean climate. There is rain in winter and hot, dry summers.

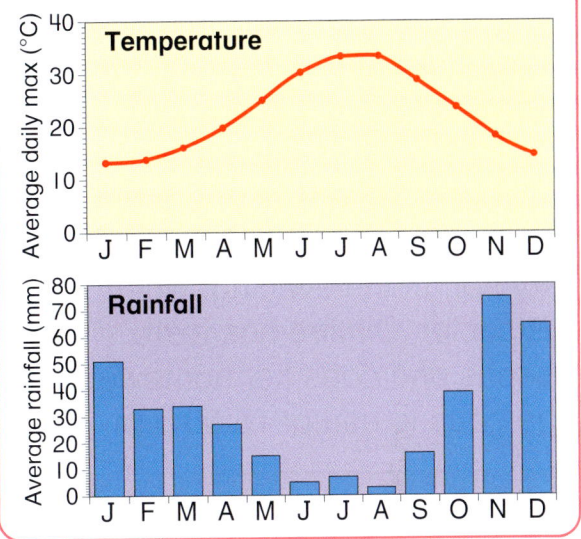

Work

Factories make metals, chemicals, clothes and electronic goods. Farming and tourism are also important.

▶ Olives are harvested for olive oil production.

Discussion

- What were the Greek people famous for in the past?
- Why is Greece such a popular place for summer holidays?
- In what ways is Greece different from your country?

Investigation

Make a fact file about Greece.

Unit 8 Greece

Lesson 2: Summer in Athens

What is the summer like in Athens?

Dimitra lives in Athens in a large flat with her family. Her school is five minutes' walk away. Lessons start at 8:00 a.m. and end at 1:00 p.m. There is a long break in the afternoon because it gets very hot. In the evening, when it is cooler, Dimitra has extra lessons and does her homework. She goes to bed at 10:00 p.m.

◀ The ferry to Amorgos.

Sleep	School	Lunch and sleep	Extra lessons	Play or watch TV	Sleep
Midnight	8:00 a.m.	1:00 p.m.	4:00 p.m.	6:00 p.m.	10:00 p.m.

▼ Buildings now surround the ruins of the ancient Greek temple in Athens.

Climate change
- Small areas of wasteland are now 'pocket parks' with grass and trees. These help cool the city and reduce air pollution.
- Better public transport and electric cars also help.

Unit 8 Greece

Each year, at the end of June, Dimitra goes to stay with her cousins on the island of Amorgos. This is a good time to leave Athens because the heat and fumes from the traffic often cause smog. Greek people call it 'nefos'. Sometimes the nefos is so bad, people are forbidden to use their cars.

Dimitra and her family travel by taxi to catch the ferry from Piraeus. At first, their journey takes them past office blocks and modern hotels. Then they go past the museum and the parliament building in Syntagma Square. The ruins of the Parthenon, an ancient temple, are in front them. Next they pass the old houses and narrow streets of the old town, or plaka. There are open-air markets and more ancient buildings.

When Dimitra last went to Amorgos, the meltemi wind was blowing in the Aegean. *"I hope the sea won't be too rough for you today,"* the taxi driver said.

Data bank
- Around 800 000 people live in Athens.
- In summer, Athens has some of the highest temperatures in Europe (well over 40 °C).
- Over 16 million tourists visit Greece and Athens every year.

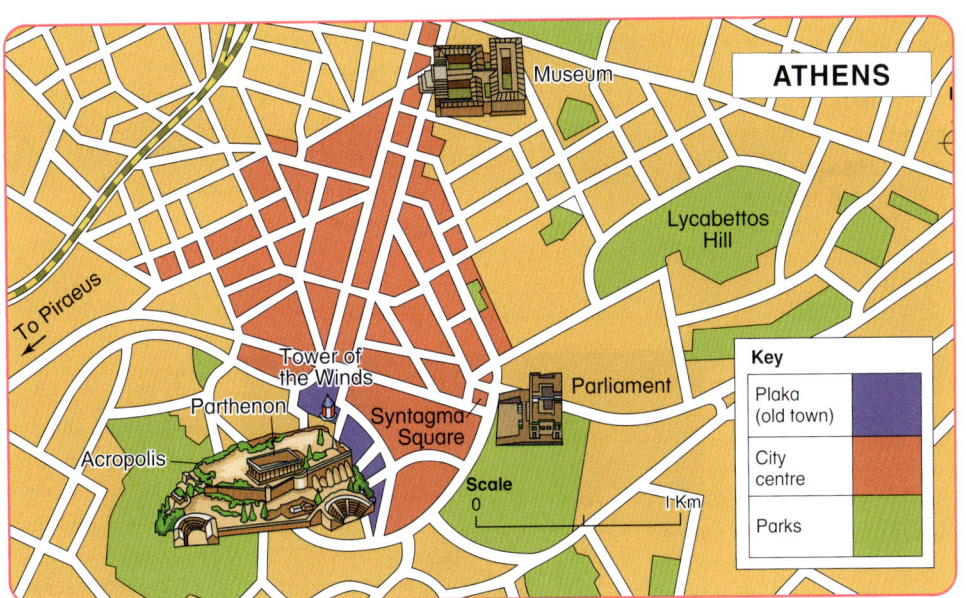

Key words
Amorgos
market
meltemi wind
smog

Mapwork
Make drawings of the things which Dimitra sees on her journey through Athens. Put them in the correct order.

Investigation
Copy the timeline of Dimitra's day. Make a timeline of your own day underneath.

Unit 8   Greece

Lesson 3: A Greek island

What is it like to visit Amorgos?

Key words

beach	taverna
monastery	route map
port	

Dimitra and her family catch the ferry from Piraeus. *"It will take about 11 hours to reach Amorgos,"* her father tells her. *"We stop at four other islands before we get there."*

Amorgos is about 35 kilometres long and 6 kilometres wide. The countryside is rocky with steep hillsides. Drivers have to be careful on the narrow road that runs from the north to the south of the island.

Dimitra's cousins live in a fishing port called Aigiali. It has a large harbour and a wide sandy beach. The town is built on a hill and has narrow streets and alleyways with steep steps.

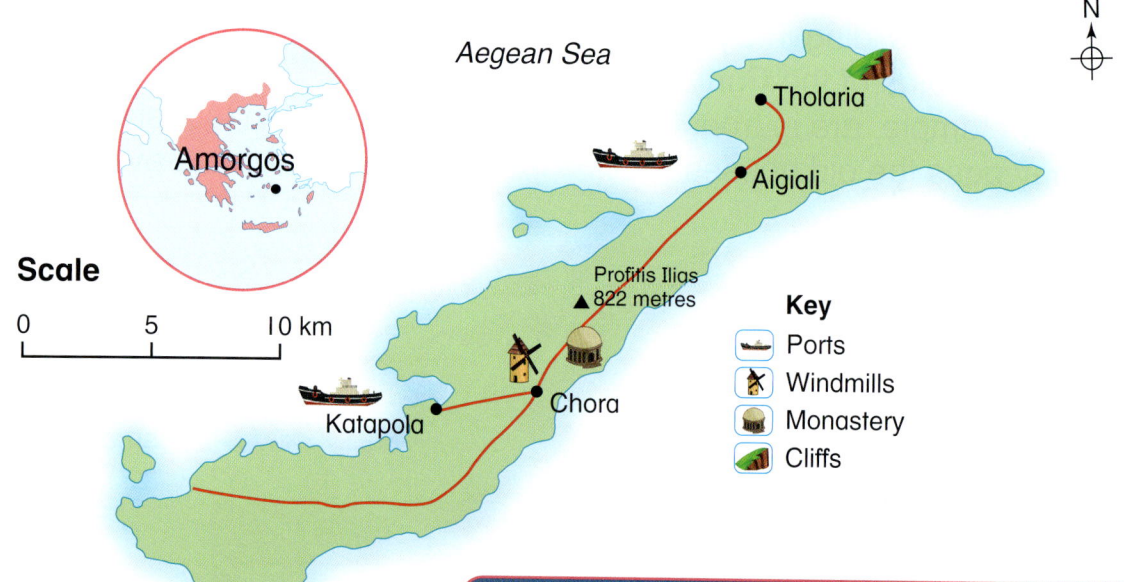

Data bank

- Ferries link Piraeus to 60 islands.
- There are usually 250 sunny days a year in the Greek islands.
- There are more than 200 islands in Greece – the largest is Crete.

▼ There were once over 30 windmills on Amorgos.

48

Unit 8 Greece

▼ Most of the houses in the town are whitewashed to keep them cool in summer.

▼ Dimitra's collection of things which remind her of the holidays.

▼ Lunch at the taverna.

Discussion
- How many islands are linked by ferry to Piraeus?
- What is Amorgos like?
- What would you most like to do if you were staying on Amorgos?

Mapwork
Draw a route map for tourists travelling from Tholaria to Katapola by bus. Describe what they might see along the way.

Investigation
Write down three things you would like to ask Dimitra about Athens and Amorgos.

Summary
In this unit you have learnt:
- about the environment of Greece
- what Athens is like in the summer
- about the Greek islands.

Unit 9 — North America

Lesson 1: Introducing the Caribbean

What is the Caribbean like?

The Caribbean consists of the Caribbean Sea, its islands and surrounding coasts. There are over 700 islands and reefs in the Caribbean and it is famous for its tropical beaches.

There are 13 independent nations in the Caribbean, including Jamaica, and Trinidad and Tobago. For hundreds of years, these islands were British colonies, while other islands were colonised by France or Spain.

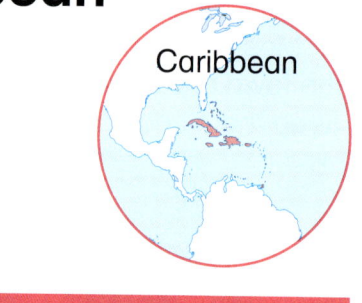
Caribbean

Key words	
hurricane	Trinidad
Cuba	Tropic of Cancer
Jamaica	Tropic of Capricorn

Discussion
- How many islands are there in the Caribbean?
- How is the Caribbean linked to Europe?
- What do you think you might like about the Caribbean?

Key
- Over 500 metres
- 200–500 metres
- 0–200 metres

Unit 9 North America

Landscape

There are two main chains of islands – the Greater and Lesser Antilles.

▲ The Pitons in St Lucia were created by volcanoes.

Climate

The Caribbean has a tropical climate. This means that there is a wet season and a dry season, and that it is warm throughout the year.

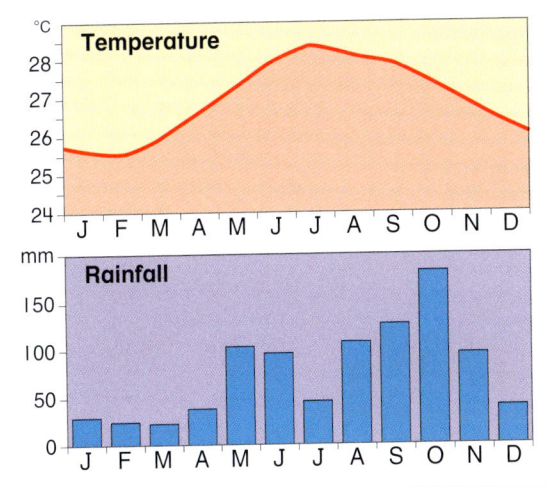

Culture

There are six official languages spoken in the Caribbean and many more unofficial languages. Among the official languages are Dutch, English, French, Haitian Creole, Papiamento, and Spanish. The Caribbean is famous for rap, reggae and other types of music.

Environment

Large areas of rainforest have been cleared for farming. Around the coast mangrove swamps and coral reefs have been damaged.

▲ From July to November there is a danger of hurricanes.

Tourism

Tourism provides jobs for many Caribbean islands. Cruise ships bring visitors from the US and Europe. Some people have private yachts.

Mapwork

Working from an atlas, make a list of islands which are on or near (a) the Tropic of Cancer (b) the Tropic of Capricorn.

Investigation

Find out about hurricanes, how they form and the routes that they take.

Unit 9 North America

Lesson 2: Finding out about Jamaica

What is Jamaica like?

Jamaica is an island in the Caribbean Sea. It is about half the size of Wales, which you studied in Unit 7. Around the coast there are sandy beaches and palm trees. The highest point is Blue Mountain Peak which is 2258 metres high.

Jamaica has a hot climate. There are only two seasons.

From November until April, it is fairly dry; from May to October it is much wetter. The mountains protect some places from the rain. Sugar, coffee and tobacco are important crops in Jamaica. The country also has valuable supplies of bauxite – the raw material for aluminium.

▼ Satellite image of Jamaica.

Mapwork

Make a map and write notes about one other Caribbean island of your choice for a class display.

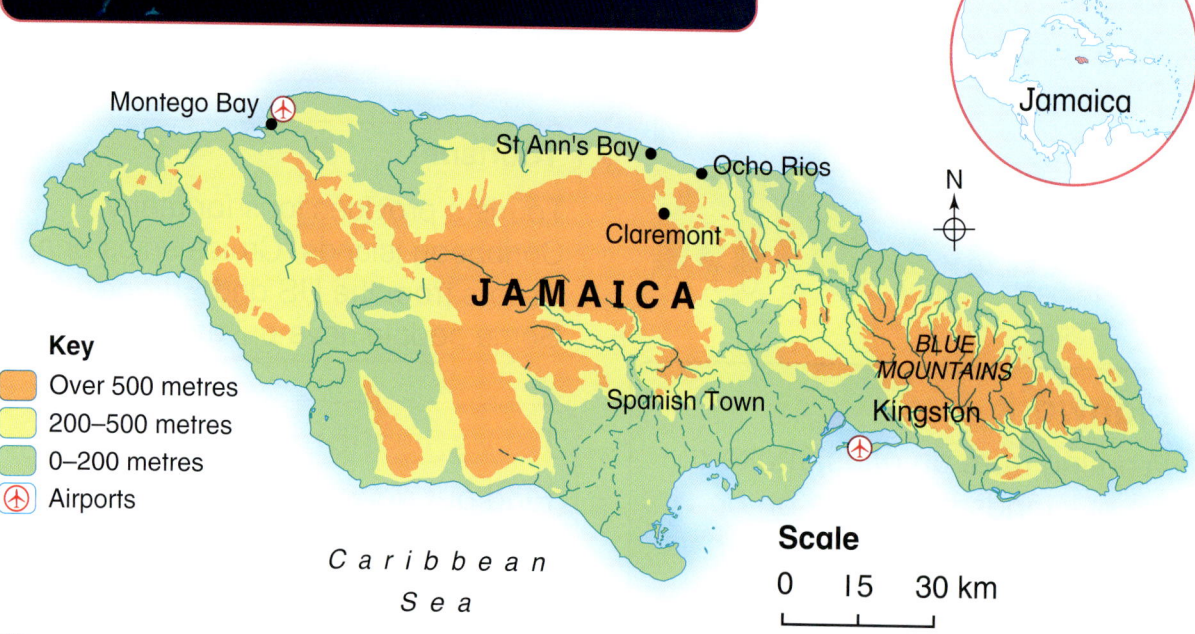

52

Unit 9 North America

Lesson 3: Living in Jamaica

How is Jamaica changing?

Ingrid Morrison is a teacher. She was brought up in Claremont, Jamaica but has lived in England for many years. The children from Riverside Primary School asked Ingrid to talk about her childhood.

▼ There are trees all around the bungalow.

"We lived in a bungalow with a verandah in the front. At the back there was an orchard where we kept chickens, cows and goats.

▼ A field of sugar cane.

In the fields all around us people grew sugar cane, bananas, oranges and lots of other fruits. They also grew vegetables like yams, sweet potatoes and peppers.

School started at 7:30 in the morning so we could have our lessons before it got too hot. I wore a tunic and white blouse for school. The boys wore khaki trousers and white shirts."

Investigation

Write a few sentences about some of the different ways Jamaica is linked to the UK through trade, sport, music and history.

Unit 9 — North America

Sunday in Jamaica

Key words

bungalow	habitats
palm trees	mangrove forest
sugar cane	parrot
sweet potatoes	seashore
verandah	

"Sunday was the best day of the week in Jamaica. In the morning people put on their smartest clothes and went to church. The women and girls all wore hats and the men put on their suits.

After the church service we had dinner. We had rice, beans, chicken and salad and a cool fruit drink with ice cream. When I was a child I used to like this better than anything else.

After dinner people played cricket or sat around talking and listening to music. Sometimes we all went to the beach at St Ann's Bay. When I was young I used to look for new shells for my collection."

▲ Women singing in the church choir.

Discussion

- What made Sunday special for Ingrid?
- What food did Ingrid's family get from the land around the bungalow?

▼ A Jamaican beach.

Unit 9 — North America

Jamaica today

"Claremont has changed quite a bit since I was young. There are many more houses now. Also people have more possessions like mobile phones and computers.

Tourists from Europe and North America come to Jamaica for their holidays. They fly to the airport at Kingston or Montego Bay and stay at hotels along the coast. The tourists bring money but the seashore is being damaged. Mangrove forests are cleared away to make room for new buildings. Parrots and other animals which live in the forests are losing their habitats.

▲ Mangrove trees have roots which come out of the water at low tide.

When I first came to England everything seemed very different. I found the weather very cold. English people stay indoors much more. In Jamaica we are always going outside and meeting our friends. I have got used to it now."

Investigation
Make a fact file of six things you have learnt about Jamaica.

▲ Today yellow-naped amazon parrots are very rare in Jamaica.

In the towns many people work in clothes factories. Some people cannot find work whereas others have plenty of money.

Summary
In this unit you have learnt:
- about the landscape of Central and North America
- about everyday life in Jamaica
- about how people can tell you about a place.

Unit 10 Africa

Lesson 1: Introducing Africa

What is Africa like?

Africa is the second largest continent. The Sahara Desert covers most of North Africa. Further south there are grasslands, rainforests and mountains. In Africa grasslands are called savannahs.

There are many great nations in Africa. In the past, these included Benin and Kongo. However, during the 19th century, European nations invaded and divided African lands between themselves to form colonies.

Most African nations regained their independence during the 1960s–1970s, and today there are nearly 60 African countries. Africa supplies the rest of the world with cheap food and metals. Cotton, coffee, tea and groundnuts are important crops. Cobalt, which is essential for batteries, comes from central Africa. Elsewhere, there are valuable supplies of copper, gold, diamonds and oil.

▲ Snow-capped Mount Kilimanjaro rises above the savannah in Kenya.

▼ Cape Town and Table Mountain at the tip of southern Africa.

Key words

Cairo	Mount Kilimanjaro
cobalt	River Nile
colony	Sahara Desert
Lake Victoria	

Discussion

- What four main types of landscape are found in Africa?
- What happened to Africa in the 19th century?
- What goods does Africa supply to the rest of the world?

Unit 10 Africa

Africa

Key
- ∧ Mountain
- Desert
- Grassland
- Forest

Scale
0   1000   2000 km

Data bank
- The Sahara Desert is the largest desert in the world (about 8.5 million km² in area).
- The Nile, Congo, Niger and Zambezi are the main rivers in Africa.
- Lake Victoria is the largest lake.

Mapwork
Using an atlas, name a country in each of the four main landscape types.

Investigation
Make up ten questions for a quiz about Africa.

▼ There are many gold mines around Johannesburg.

57

Unit 10   Africa

Lesson 2: Learning about Kenya

Finding out about Kenya

The landscape of Kenya is one of the most varied in Africa. There are deserts, farmlands, savannahs, rainforests and extinct volcanoes. The coast has beautiful beaches fringed with palm trees.

Most people in Kenya live in the central highlands where the cool climate is good for farming. Coffee, tea, maize, fruit and vegetables are important crops.

Kenya was once a British colony, but regained its independence in 1964. English and Swahili are the main languages. The capital, Nairobi, is growing fast.

▲ Tourists come to Kenya to visit the game parks.

▼ Nairobi

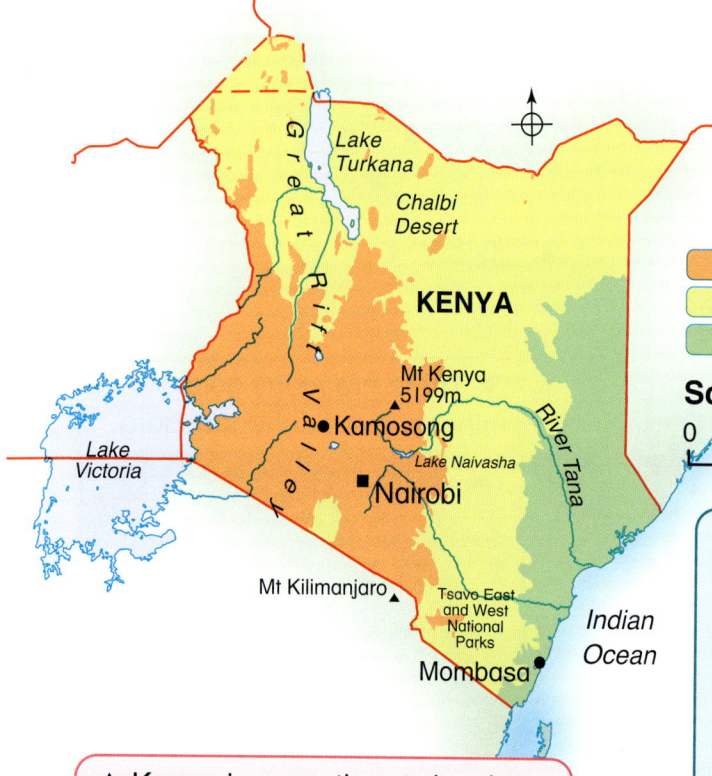

▲ Kenya is more than twice the size of the United Kingdom. It has a population of around 55 million.

Key words

central highlands
extinct volcano
Kamosong
pyrethrum
savannah
Swahili

Discussion

- How would you describe the landscape of Kenya?
- What do farms produce in Kenya?
- What would you find different if you lived with Miriam's family?

Unit 10 Africa

Making links with a school in Kenya

At St Peter's School pupils wanted to find out more about Kenya. Their class teacher set up a link with a school in a village called Kamosong about 300 kilometres from Nairobi. The children told each other about the place where they lived. They also drew maps of their local area.

▲ Miriam

The children made up fact files of the information they found out.

Where does Miriam live?

- Miriam lives near the city of Eldoret in the Rift Valley region of Kenya.

- Miriam's home village is called Kamosong. She lives around 5 km from the centre.

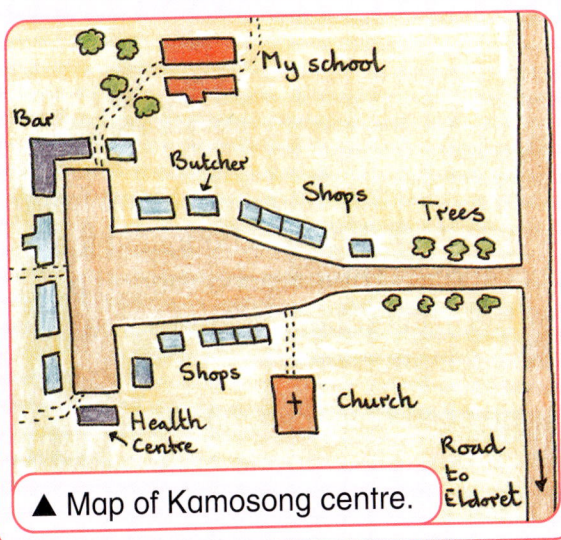
▲ Map of Kamosong centre.

Miriam's home and family

- Miriam has four brothers and three sisters. There are eight children in the family.
- The family lives together in a house made of earth and thatched with grass.

School

- Miriam goes to school at Kamosong Primary school.
- She gets up at 6:00 a.m. to go to school.
- Her favourite subjects are Art, Swahili and Science.

Work

- Miriam grows pyrethrum which is used to make insect repellent. She weeds and looks after the plants when she is not at school.

▲ Pyrethrum flowers.

Hobbies

- Miriam likes drawing, taking photos, reading and making friends.
- She goes to Girl Guides once a week.

Investigation

Write a description of Miriam's life using these headings: (a) family (b) hobbies (c) school day.

Unit 10 Africa

A parcel from Kenya

Investigation
Select six items from the display table. Write a sentence saying what each one tells you about Kenya.

1. A packet of coffee.
2. A shaker made from bundles of reeds.
3. A strip of cloth with traditional patterns.
4. A newspaper with headlines about a local rise in taxi fares.
5. Kenyan bank notes.
6. A sandal bought in the local market made from a recycled car tyre.
7. Green beans grown in Kenya and sold in shops in the UK.
8. Photographs of the school and houses where the children live.
9. Postcards of animals in one of the safari parks.
10. Water pots.
11. A children's book.
12. A map showing places to visit around Kamosong.

One day the children received a parcel from their friends in Kamosong. It contained a selection of newspapers and postcards, a few bus tickets, a strip of cloth and some photographs which the children had taken of each other. There were also some interesting stamps on the outside of the parcel.

The children arranged everything on a display table. Some of the class added fresh fruit and tins of food from Kenya which they had bought in the local shops. Their teacher added a few Kenyan ornaments and some musical instruments.

As the children continued with their project, they found other sources of information. They read books about Kenya and watched a television programme.

Discussion
- What is your favourite object in the display?
- What would you like to investigate about Kenya?
- Which six things would you put in a box to describe your own life?

Unit 10 Africa

Lesson 3: Living in Kenya

How is Kenya changing?

Key words	
annotated map drought	game park

Here are some of the things which the children at St Peter's found out about life in Kenya today.

Climate change
In recent years the rainfall has been unreliable. Heavy rain has caused floods while other regions have suffered droughts.

Game parks
There are over 15 game parks in Kenya. These are popular with tourists but local people are sometimes forced out of areas to make more space for animals.

◀ Droughts are causing deserts to spread.

Farming
Some people earn money selling beans and roses to Europe. These plants take valuable water which is needed to grow crops for local people.

Cities
Nairobi, Mombasa and other cities are growing larger. People are moving there in search of work.

▲ Aeroplanes bring roses from Lake Naivasha for sale in the UK.

Mapwork
Draw a map of Kenya. Annotate it with notes round the edge saying how Kenya is changing.

Summary
In this unit you have learnt:
- about the cities and countryside of Africa
- how links with another school can help provide information
- about some of the issues in Kenya today.

Investigation
Design an advertisement to attract tourists to a Kenyan game park. Include information about its location, the weather and the journey from your country.

Glossary

Bulk carrier
A lorry designed to carry heavy goods like sand and gravel.

Bungalow
A type of house, originally from India, often one storey high with a verandah.

Carbon emissions/ CO_2 emissions
Climate changing pollution caused by burning fuels like gas and oil.

City
A very large settlement. In the UK, cities usually have a cathedral.

Channel
The route taken by water in a stream or river.

Climate
The pattern of weather over many years.

Cobalt
A hard silvery metal.

Colony
A country which has been taken over and exploited by another country.

Crop
Plants which farmers grow and harvest such as wheat, apples or bananas.

Extinct volcano
A volcano which was once active but is now completely dead.

Fish stocks
The quantity of fish in the world.

Fumes
Harmful gases which damage the health of people, plants and animals.

Harbour
A sheltered area on the coast for boats and shipping.

Hurricane
Violent storms which form in tropical oceans and do great damage when they reach land.

Magnetic compass
An instrument with a suspended magnetic needle that points to the North and South Poles.

Mangrove forest
Forests that grow on the seashore in the Tropics.

Minerals
Materials such as oil, gold and silver which are found naturally in rocks and soils.

Nuclear waste
Radioactive material which is left from industrial activity, for example, making electricity.

Oil platform
A large structure for drilling wells and extracting oil at sea.

Ranging pole
Red and white striped pole for measuring slopes and changes in level.

Resource
Something which people find useful.

Settlement
The places where people live, such as villages, towns and cities.

Skyscraper
A very tall building, often built round a steel frame and used as flats (apartments) or offices.

Suburbs
The outer edges of a town or city where there are often houses and schools.

Tropics
Parts of the world where the sun is directly overhead at least once a year.

Turbine
A machine with blades that generates power.

Index

Aegean Sea 44
Africa 56–61
Amorgos, Greece 46, 47, 48, 49
Anglesey, Wales 39
animals 2, 3, 6, 13, 14, 55, 60–61
Antarctica 4
Arctic Ocean 4, 5
Athens, Greece 44, 45, 46, 47
Atlantic Ocean 4

Big Pit, Wales 40–43
birds 7, 14
Blaenavon, Wales 40, 41

Cape Town 56
carbon 34
Caribbean 50, 51, 52, 53
Cardiff, Wales 38, 39
cargo 5, 29
cheese farming 18
chemicals 30, 33, 35, 45
cities 20–25, 61
city populations 22
climate 16, 17, 34, 45, 51, 52, 58
climate change 4, 11, 18, 20, 25, 35, 46, 61
coal 5, 35, 36, 38, 40–41, 42, 43
cod 4
colonialism 50, 56
coral reef 2, 4, 6, 51
Corinth Canal, Greece 45
cows 18, 53
crops 14, 16, 17, 52, 53, 56, 58, 61
cut-offs 11

dams 11, 35
Dart, river 12
deposition 9
Doha, Qatar 20
Drax power station, UK 36
dykes 11

electricity 35, 36
energy saving 34–35
environment 5, 12, 32, 33, 34, 35, 36, 37, 51
equator 4
erosion 9
Eryri (Snowdonia) 39
Europe 16, 24
European Union 41, 44

factory 10, 26, 27, 28, 33, 34, 38, 39, 41, 45, 55
farming 10, 18, 28, 30, 39, 45, 51, 58, 61
ferries 39, 45, 46, 47, 48
fish 2, 3, 6
fishing 4, 7, 12, 30

floods 10, 11, 22, 61
fruit 14, 33, 53
fumes 21, 32, 33, 34, 36, 37

game park 57, 60
gas 5, 6, 34
gas platforms 6
Greece 43–49

herring 3
high-speed trains 20
Hong Kong 4, 20

Jamaica 51–55
jobs 25–31
Johannesburg 56

Kamosong, Kenya 58–59
Kenya 57, 58, 59, 60, 61

Lake Naivasha 60
lakes 8, 9, 56
levees 10
local investigations 12, 36, 59
London, England 23–25

machines 25, 27
maps 4–5, 7, 10, 12, 16, 17, 22, 25, 38, 41, 44, 47, 48, 50, 52, 57, 58, 59
Marianas Trench 3
Mediterranean 16, 44, 45
minerals 2, 6
mining 38, 40, 41, 42, 43, 57
Mississippi, river 10–11
monsoon 17
motorways 21, 39
Mount Everest 3
Mount Kilimanjaro 56, 57

Nairobi, Kenya 58, 59, 61
New York, US 23
Nile, river 57
noise 32, 37
Normans 24
North America 50–55
Northern Wolffish 3
North Sea 7, 19

ocean currents 4
oceans 2–5
offices 21
oil 5, 6, 7
oil platforms 6, 7
oil spill 34

Pacific Ocean 4, 10
Pindus Mountains, Greece 45
plants 14, 61
pollution 5, 7, 32–37

railways 40
rain 10, 11, 17, 19, 39, 45, 51, 61
recycling 20, 26, 34–35
renewable energy 35, 36
reservoirs 9
rivers 8–13, 24, 33, 35, 39
roads 21, 45
Romans 24

Sahara Desert 56–57
scientists 2, 5
seas 6–7
seaside resorts 6, 19, 39
seasons 14–19, 52
sheep 18, 39
ships 5, 7, 10, 24, 51
shops 21
skill 28
skyscrapers 20–21
snow 19, 23, 39
solar panels 35
Southeast Asia 17
storms 4
streams 9
submarine 2, 3, 5
suburbs 21
Switzerland 18

Tacoma Building, Chicago 21
tankers 5
Thames, river 24
tourism 19, 38, 44, 45, 47, 51, 55, 58, 61
trains 21, 38
transport 5, 39, 45, 46
transportation 9
turbines 35, 36

underground trains 24, 25
underwater vents 2
United Kingdom 7, 16, 38–43
United States 10, 23

volcanoes 3, 32, 51, 58

Wales 38–43
water cycle 8–9
weather 2, 4, 14, 16, 17, 18, 39
whales 4
Whitby, Yorkshire, England 19
wind 19, 35, 39
wind farms 7, 35
windmill 48
work 39, 45
work force 30
Wye, river 39

63

William Collins' dream of knowledge for all began with the publication of his first book in 1819.

A self-educated mill worker, he not only enriched millions of lives, but also founded a flourishing publishing house. Today, staying true to this spirit, Collins books are packed with inspiration, innovation and practical expertise. They place you at the centre of a world of possibility and give you exactly what you need to explore it.

Published by Collins
An imprint of HarperCollins*Publishers*
The News Building, 1 London Bridge Street, London, SE1 9GF, UK

HarperCollins*Publishers*
Macken House, 39/40 Mayor Street Upper, Dublin 1, D01 C9W8, Ireland

Browse the complete Collins catalogue at collins.co.uk

© HarperCollinsPublishers Limited 2025

Maps © Collins Bartholomew 2025

10 9 8 7 6 5 4 3 2

ISBN 978-0-00-872832-8

All rights reserved. No part of this publication may be reproduced, stored in a retrieval system, or transmitted in any form or by any means, electronic, mechanical, photocopying, recording or otherwise, without the prior written permission of the Publisher or a licence permitting restricted copying in the United Kingdom issued by the Copyright Licensing Agency Ltd, 5th Floor, Shackleton House, 4 Battle Bridge Lane, London SE1 2HX.

Without limiting the exclusive rights of any author, contributor or the publisher, any unauthorised use of this publication to train generative artificial intelligence (AI) technologies is expressly prohibited. HarperCollins also exercise their rights under Article 4(3) of the Digital Single Market Directive 2019/790 and expressly reserve this publication from the text and data mining exception.

British Library Cataloguing-in-Publication Data

A catalogue record for this publication is available from the British Library.

Authors: Stephen Scoffham and Colin Bridge (with additional original input from by Terry Jewson)
Publisher: Laura White
Product manager: Natasha Paul
Development editor: Judith Walters
Copyeditor and proofreader: Catherine Dakin
Cover designer and illustrator: Steve Evans
Internal illustrators: Jouve India Private Ltd and Hannah Drennan, Beehive Illustration
Typesetter: David Jimenez
Production controller: Alhady Ali
Printed and bound in the UK by Bell and Bain Ltd, Glasgow

This book is produced from independently certified FSC™ paper to ensure responsible forest management.

For more information visit: www.harpercollins.co.uk/green collins.co.uk/sustainability

Acknowledgements

The publishers gratefully acknowledge the permission granted to reproduce the copyright material in this book. Every effort has been made to trace copyright holders and to obtain their permission for the use of copyright material. The publishers will gladly receive any information enabling them to rectify any error or omission at the first opportunity.

P13t Richard Allaway; P16 an adapted extract from *Memed My Hawk* by Yashar Kemal, translated into English by Edouard Roditi, Harvill Press, copyright © Yashar Kemal, 1955, 1958, translation copyright © Edouard Roditi, 1961. Reproduced by permission of The Random House Group Limited; and Aragi Inc. on behalf of the Yashar Kemal Estate. All rights reserved; P17 An adapted extract from *Plain Tales* from the Raj by Charles Allen, Little Brown, copyright © Charles Allen, 1975. Reproduced by permission of Sheil Land Associates Ltd; P40bl, p40br © National Museum Wales; P41 hand drawn map of Blaenavon, copyright © Torfaen County Borough Council Blaenavon World Heritage Site. Reproduced by permission of Visit Blaenavon; P42 Jeff Morgan 03/Alamy Stock Photo; P49tr, P60 © Stephen Scoffham.

All other photos Shutterstock.